ENCYCLOPAEDIA OF NANOTECHNOLOGY-V

MOLECULAR ENGINEERING

By

Dr. M.P. Arora

Dept. of Zoology
M.M.H. Post Graduate College
Ghaziabad
(U.P.)

DISCOVERY PUBLISHING HOUSE PVT. LTD.
NEW DELHI-110 002

First Published-2008

ISBN 978-81-8356-281-2

Published by

DISCOVERY PUBLISHING HOUSE PVT. LTD.
4831/24, Ansari Road, Prahlad Street,
Darya Ganj, New Delhi-110002 (India)
Phone: 23279245 • Fax: 91-11-23253475
E-mail: dphbooks@rediffmail.com
dphtemp@indiatimes.com

Printed at:

Sachin Printers, Delhi

Preface

Everyone in the modern world knows what technology is. But what is nanotechnology? Taken from the Greek, *nano* means "one billionth part of" a whole. In modern parlance, it means very, very small. In today's world of smaller cell phones and portable computers, miniaturization is the current technology. Nanotechnology is the next step after miniaturization. In tomorrow's world, nanotechnology will be the new common technology. It will affect everyone of the planet and many change civilization as we know it.

Nanotechnology is the molecular engineering. It will soon create effective machines as small as DNA. It describes the ideas and techniques that are creating or new domain of science and technology. Nanotechnology holds the promise of revolutionizing field ranging from medical science to optical communications, but requires a substantial investment in time and money to make the promise become a readily. The results could be nothing less than a new industrial revolution.

In an undertaking as new as nanotechnology, many different approaches are studied in order to ascertain the best way to fashion working devices. At present, techniques can be divided into one of two general categories. One technique, termed the top-down method, physically manipulates molecules into nanostructures. The other technique, termed the bottom-up method, coaxes atoms and molecules to assemble themselves into nanostructures.

The present title **"Molecular Engineering"** has been designed for undergraduate and post-graduate students as well as those involved in basic research in biological, biochemical and biophysical sciences.

In the preparation of this book large number of books and research papers have been consulted. So no authenticity is claimed.

The author expresses his gratitude to Mr. Wasan and staff of M/s Discovery Publishing House for their whole hearted co-operation in the publication of this book.

The author tried hard to be accurate and upto date in statement and realises the impossibility of completely avoiding errors therefore, the author will greatly appreciate having his attention called to any questionable statement.

Author

Contents

1

INTRODUCTION

Molecular nanotechnology is basically the branch of experimental applied science that will focus on conceptual issues that show how molecular nanotechnology relates to other technologies, how it fits in with our understanding of science and of existing engineering practice, and how it will work in a basic physical sense. There is a clearcut distinction between pure and applied science, and also the distinction between theoretical and experimental science. What is not so often noticed is that these distinctions are orthogonal.

In pure theoretical science, for example, people propose grand unified theories of physics. Similarly in pure experimental science, the people, for example, try to test those theories using high-energy particle accelerators. There is also a tremendous amount of effort put into experimental applied science. People work in the laboratory to try to make things that are useful.

Molecular nanotechnology will become an experimental applied science, and then mature into a branch of engineering. Today, however, it is chiefly a theoretical applied science. It is "theoretical" in that one is doing not experiments, but computational modeling and theoretical analysis. Yet, it is "applied" in that the systems being studied and modeled are not phenomena of nature or entertaining objects of academic interest, but are instead designs for artifacts intended to be useful, once built. Many of these will be manufacturing systems.

Theoretical applied science does not have the prestige of pure science, because it is concerned with building things that will be useful to people, nor does it have the near-term economic value of experimental science, because it studies systems that cannot be built quite yet. Since people are usually paid for producing results in the short term, there is remarkably little money available for this kind of work. So it has been, to an amazing extent, unexplored territory.

Essential Role of Manufacturing

The theoretical applied science draws broad conclusions about technology, it is natural to start with manufacturing, which is how technological products are made. Manufactured products are made from atoms, and their properties depend on how those atoms are arranged. For example, coal can be converted into diamond by rearranging its carbon atoms. Likewise, silicon atoms extracted from sand can be rearranged to make semiconductor chips.

This observation points to a basic shortcoming of our technology. The world is made of atomic and molecular building blocks, but today we lack anything like a hand able to place those fundamental building blocks in chosen patterns. Chemists today make complex molecular structures by taking smaller pieces, putting them together, stirring, and hoping that they will fall together to make the right product. If you imagine trying to make an automobile by taking parts, putting them into a box, shaking, and hoping that they will fall together to make a working machine, you will conclude that it is very useful to have robots, or hands, or something like them involved in the process.

Chemists have achieved remarkable things, given that they have no *molecular hands* with which to put the parts where they want them. It is amazing that chemists are able to do anything at all, and, in fact, they have impressive and growing accomplishments.

This conclusion about the importance of hands points to a basic technological opportunity. Molecular "hands" can be developed, and can be used to build more and better molecular hands. Some consideration of how molecular hands can be developed is included here, although the focus of this presentation will be on what will become possible with advanced systems near the culmination of the process of building better molecular hands.

Early Ideas and Recent Progress

Building with the fundamental building blocks of nature is not a new idea. It was stated by Richard Feynman who in 1959 draw the attention. The principles of physics, do not speak against the possibility of maneuvering things atom by atom. Ultimately chemical synthesis is brought about put the atoms down where the chemist says, and so you make the substance. No one has seriously challenged this view, although many have been skeptical of its long-term implications. One good reason for not challenging this view is work that was published in the journal *Nature*. The surface of a nickel crystal covered by white dots that are, in fact, xenon atoms. The atoms were imaged by a *scanning tunneling microscope* (STM), which moves a sharp tip near the surface with atomic precision. This picture graphically demonstrates that, yes, atoms are, as we were taught in high school, little fuzzy balls that have mass, occupy a position, and can be moved around. In this case, 35 atoms were placed to spell "IBM."

Assuming that the underlying nickel crystal is a precise structure, this is an atomically precise structure. Each of these xenon atoms is like an egg sitting in a cup in an egg crate. It is either in the right hole or in the wrong hole. There is no stable intermediate position. The positions of the atoms are like states in digital logic, in which every signal is either "1" or "0." It is either exactly right or exactly wrong (at least as seen by other logic devices). If the system is structured correctly, the probability of its being exactly right can be made extremely high. In this case, researchers fiddled until all 35 atoms were in the right places. The probability of getting an atom in the right place on the first try was low and it took about a day to make the IBM logo. But it was possible to keep fiddling until everything was right. If you had a better mechanism than this STM, then you could get many operations right the first time, with high reliability. There is a high level of reliability in digital logic, where millions of instructions per second are performed by a computer for days on end, and each of the instructions is done right.

This *atomic-scale logo* suggests the precise construction at the molecular scale. One problem with extending the techniques used here is that the IBM logo was built using inert gas atoms at a few degrees Kelvin. At room temperature, such a structure would not be stable. Stable atomic-scale modifications have subsequently been made by broadly analogous techniques, but one would like to have a more general ability to make things. One would like a device more like a hand that was able to hold a piece and put it where you wanted it, where that piece would now be a reactive molecule that would bond to another molecular structure. The products would be stable at room temperature, and could serve as parts in a system of molecular machinery.

Implications of Atomic-scale Positional Control

An atomic force microscope (AFM) is commonly used in experimental work in the branch of nanotechnology. An STM senses

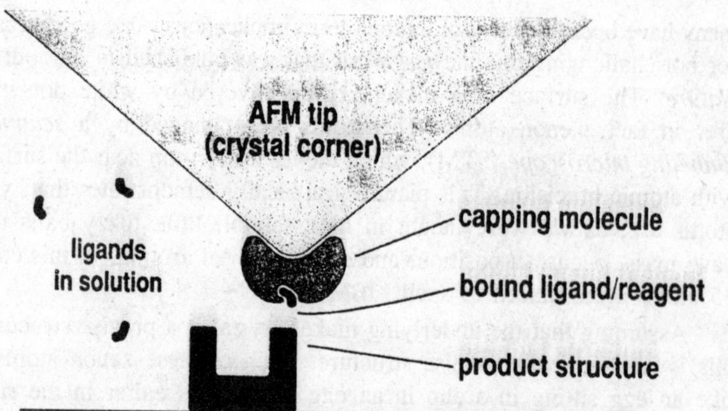

Fig. 1.1. A molecular manipulator using a protein bound to the tip of an AFM to maneuver small ligand molecules.

its proximity to a surface by the time ling current produced by electrons jumping the gap between the tip and surface. The AFM, however, touches the surface and senses contact with the surface by measuring the deflection of the tip. (However, it must be admitted that the difference between a gap and a contact is a bit fuzzy, since atoms are fuzzy on their surfaces.) The AFM tip is shown schematically, with a protein molecule bound to the tip. The protein, in turn, is binding a ligand, a small reagent molecule. As you move the tip around, you move both the protein "gripper" and the reagent bound to the "gripper." The tip can, therefore, be used to bind reagent molecules from solution, hold one molecule with high probability at that point, and then move that molecule to induce a chemical reaction at a particular location.

The implications of atomic-scale positional control for the manufacture of complex molecular structures can be appreciated from the following analysis. Standard models for chemical reactions relate the rates of reactions to the effective concentrations of the reacting molecules. Positional control of reactants can produce very high effective concentrations at a chosen location and low concentrations elsewhere. To estimate the magnitude of this effect, it is necessary to know how precisely a reactant could be positioned with an AFM.

An analysis of the extent to which thermal motions displace a molecule held by an AFM tip is made. This analysis describes the cloud of probability surrounding the point at which you want the molecule to be, giving the probability of finding the molecule near each point within that cloud. This probability density describes the effective concentration of a suitably oriented molecule, and it can be as high as 150 molar for reasonable parameters of the system. This result takes account of laboratory vibration, which adds to the uncertainty caused by thermal vibration. The conclusion is that the concentration of the molecule where you want it can be 10^8 times higher than at adjacent points, only an atomic diameter away. This conclusion is based on the properties of existing AFM mechanisms and proteins. An AFM with a gripper is something we have not yet seen, but which I believe we will see in the relatively near future.

This idea of an AFM-based molecular gripper has stirred some interest, and there are now several groups interested in moving in this direction. Dr. Shin-ichi Kamei has decided to explore the idea of molecular nanotechnology based upon molecular LEGO and an AFM tip with a binding protein. As a further example of growing interest in this general area, the journal *Nature* published (August 22, 1991) an announcement of the intentions of Japan's Ministry of International Trade and Industry (MITI), which "has decided to launch a major government industry initiative to develop nanotechnology." MITI plans to spend about $185 million over 10 years. That adds to funds that were already flowing into this field, from the Science and Technology Agency of Japan.

Uses of the Term "Nanotechnology"

There is still some ambiguity in what is meant by "*nanotechnology*." Nanotechnology based on mechanical control of chemical synthesis is only one of a number of uses of the term.

After the First Foresight Research Conference on Molecular Nanotechnology, Science News carried an article titled "Nonexistent Technology Gets a Hearing." The article said that, yes, a technology based on molecular machinery building additional molecular machinery does not yet exist, but that various researchers have concluded on the basis of analysis and analogy with biology, and so forth, that such a technology can be developed, and that it is of consider able importance today to understand where we are going, how to proceed, and what it means for broader concerns. News about the most recent achievement in "nanotechnology" are being published. It exists now, and what it is about is melding submicron particles by precipitation of minerals in a gel. People have been doing things like that for thousands of years, whether they knew it or not. Thus, the existing fields of activity relabeled as "nanotechnology" because it has become a buzzword.

Let us make some distinctions in terminology, starting with *micromachines*. One sees many pictures of micromachines in the press, but no one has yet shown how micro machines are particularly relevant to the development of molecular nanotechnologies. *Micromachines* are quite appropriately an active area of research because they have many potential uses. But, in my assessment, if the development of micromachines were to stop dead today, it would make little difference to the development of molecular nanotechnology. Micro machines thus seem essentially irrelevant to molecular nanotechnology, although uses may turn up that are not yet recognized.

Molecular nanotechnology refers to complex, atomically precise functional structures. The word "nanotechnology" without the modifier "molecular" has become a buzzword that is often applied to glamorize the production of nanometer-scale blobs. Saying this is not to judge these latter efforts harshly. Microelectronics and personal computers are made possible by the ability to make micron-scale blobs, so that ability is very useful. Making such blobs even smaller will be even more useful. This kind of "nanotechnology" (without the modifier "molecular") is, in part, about taking the familiar, ongoing, enormously important revolution of microelectronics, and carrying it further. This is wonderful, but it is not molecular nanotechnology. It is an extension of micro technology.

Molecular manufacturing is also called *molecular fabrication* because these terms clearly describe the long-term goal. These terms are also long and inconvenient enough that perhaps researchers will not appropriate them to describe things that they are doing today. The basis of molecular manufacturing is "*mechanosynthesis*," the use of mechanical control to guide the placement of molecules so as to build complex objects. "Molecular machine systems" are the chief enabling technology and the chief intermediate goal in developing molecular nanotechnology.

There are two basic paths to molecular manufacturing. Molecular machine systems can be developed, use them to perform *mechanosynthesis* (including the mechanosynthesis of more molecular machine systems), and then use that technology base to make a wide range of products. Alternatively (as with the AFM path), we can develop the capability to do mechanosynthesis, use that to build molecular machines systems, use those to build more molecular machine systems by way of mechanosynthesis, and proceed in the same way to make a wide range of products. Therefore, mechanosynthesis and molecular machine systems are two different entry points into a spiraling technology base in which mechanosynthesis is used to make molecular machine systems that are used to perform mechanosynthesis. That spiral leads to advanced molecular manufacturing.

CONSEQUENCES OF MOLECULAR NANOTECHNOLOGY

The following discussion of the longer-term consequences of molecular nanotechnology is based on technical analyses that have only recently been published. *Nanosystems* examines molecular machinery, manufacturing, and computation (both in the sense of computational modeling and of computational applications).

The physical principles underlying the study of nanosystems have been dealt in the present volume. This part discusses scaling laws, which show, based on simple classical mechanics, that you should expect very small devices to have high frequencies of motion, high power densities, and several other desirable physical features. It also shows, for example, that electrostatic motors are a good idea on a small scale, but that electromagnetic motors are a very bad idea on a small scale. After that background, it discusses the potential energy as a function of shape for molecules, which determines strength, stiffness, dynamics, and so forth. These potential energy functions are fundamental to a molecular analysis of nanomachines.

A survey of molecular dynamics follows, concluding that classical mechanics is good enough for most of the analysis, with occasional quantum corrections and the use of statistical mechanics to describe thermal vibration. A study of positional uncertainty describes how much components are displaced by a combination of quantum and thermal effects. A study of transitions, errors, and dam age discusses how things break, how they can get put in the wrong place, and stay there. A section on energy dissipation explains how the motion of these parts converts useful energy into waste heat.

Structural components are discussed in a general sense, as well as basic moving parts, and intermediate subsystems (such as pumps and motors). In analyzes nanomechanical computations systems, describing computers able to perform 10^{16} instructions per second per watt, which is about a billion times more computation per unit power than we can do today. Because of their compactness, such computers also allow about a billion times more computers to fit in a box of given volume than is possible today. This section concludes with a pair of chapters that, taken together, describe the process of building macroscopic objects by combining molecular building blocks.

The pathways for implementing molecular nanotechnology have also been discussed. In the begining of this part developments that are today one step from the laboratory, and that are reasonable goals for the next few years. These include developments in protein engineering, in other sorts of macromolecular engineering, and in positioning molecular building blocks using AFM. The final chapter discusses paths from current and near-term technologies through a series of steps using tools to build better tools, leading finally to advanced molecular manufacturing.

Importance of Molecular Manufacturing

Why is molecular manufacturing of interest? In general, molecular manufacturing can be cleaner than existing process technologies because it involves more precise control of where molecules go and how they are rearranged. Existing processes often involve combustion and produce unwanted side products, such as nitrogen oxides and ash. Furthermore, intermediate products often must be cleaned, leaving for disposal an often toxic solvent containing other toxic materials. Molecular manufacturing methods, in contrast, can be clean because they do things right the first time and produce only the specified products.

Further, structures can be made stronger with molecular manufacturing because they can be made out of perfect graphite fiber and perfect diamond fiber composite. Such materials are approximately 70 to 80 times stronger on a per-unit mass basis than are aerospace aluminum and high-strength steel. This has tremendous consequences for performance. Products typically can be much lighter (in part because computers can be made much smaller and motors can have much higher power densities).

Molecular manufacturing also can be highly efficient. If you only need 1/80th of the material to build a structure of equivalent strength, and if the process itself is reasonably efficient, then you use less material and less energy. The combination of higher strength, lighter weight, and higher efficiency expands capabilities across the board.

There are also large increases in reliability when a typical defect is having one molecule out of place. For some systems, such a defect will be fatal because every molecule must be in the right place, but for systems like the ones we use today, defects on this scale are indistinguishable from perfection.

Finally, by using common inexpensive materials like acetone (or even carbon dioxide from the atmosphere, given sufficient energy input) as raw materials, and by processing these materials with systems that are themselves fast and efficient, the manufactured products can be inexpensive and, therefore, more available.

Molecular Machine Components

Ralph Merkle discussed atomically precise designs in more detail. The bearing has ridges on the shaft that interlock with ridges on the sleeve. Analysis shows that this structure is very stiff. If you attempt to bob the shaft up or down, or rock it from side to side, or displace it in any direction, you encounter a very strong resistance. However, the symmetry properties of the rings ensure that the shaft turns easily within the sleeve because a small angular displacement returns the energy of the system to its previous value. As the shaft turns within the sleeve, the energy of the system oscillates up and down very rapidly. Because atoms are soft and smooth, any very rapid variation in energy must be extremely small. In contrast, if the variations occurred over a distance as large as an atomic diameter, then there easily could be a large energy barrier. Theoretical analysis with the standard computational models used by chemists indicates that the energy barriers to rotation of the shaft within the sleeve are less than 1/1000th of the vibration energy of a single molecule at room temperature. Thus, the bearing is extremely stiff in all directions except axial rotation, and in that direction it is extremely smooth. With essentially no static friction, it is a good bearing. The bearing is one example of a simple molecular mechanism. It shows that we are considering rigid objects that behave like the bearings, gears, and similar components that one finds in macroscopic systems.

To do molecular manufacturing, you must start with molecules. If the inside of a molecular manufacturing system is to work by handling molecules rapidly and deterministically in much the way that bits are handled in a computer, you must have molecules of just one kind coming in through a particular port of entry. These streams of molecules must be highly purified. A schematic diagram of a device that has designed in atomic detail in a modest length of time on a graphics workstation. It shows a rotor that surrounds a cam, with the whole structure embedded in a wall. The rotor includes sliding rods; when the rods are retracted, the pockets on the surface of the rotor form good binding sites for a particular molecule. There are other molecules outside in solution, but only one kind fits the pocket nicely. As the rotor turns, these pockets carry the correct molecules into the interior. The cam surface thrusts out the sliding rod, emptying and blocking the binding site as the sites move back toward the outside. Thus, as the rotor turns, it selectively pumps specific molecules from one side of the rotor to the other.

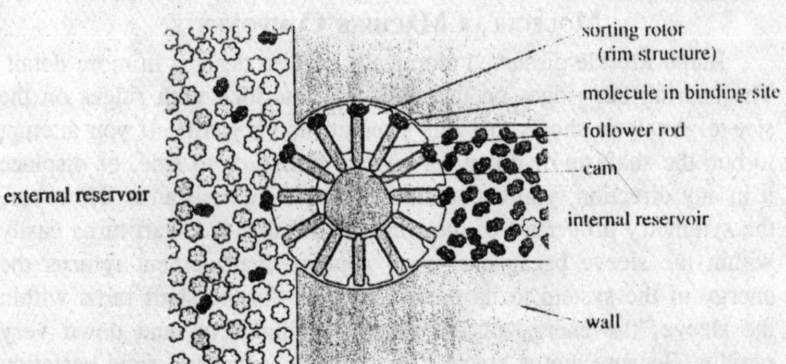

Fig. 1.2. Selective molecule pump based on a sorting rotor with modulated receptors.

People occasionally have objected to this pump because they imagine that it would be an example of a Maxwell's demon. The answer to that objection is that you must put power into the system to turn the shaft. Pumping these molecules typically increases the pressure, and it certainly increases the concentration of this species. Therefore, this pump is doing work and must be connected to a motor.

A series of these pumps in a staged cascade can transport molecules from a solution that contains one percent of what you want and 99 percent of what you do not want, to a solution that contains almost exclusively what you want, and only one molecule in 10^{15}, that you do not want. With one impurity molecule in 10^{15}, you can watch molecules going by at a million per second for about 30 years before seeing one of the wrong kind. This can provide a basis for a highly deterministic process in the interior of a molecular manufacturing system.

Molecular Manufacturing System

To picture a molecular manufacturing system, start by picturing a box surrounded by a wall. Raw materials from a tank are taken in through the wall, and what occurs inside the box is precisely controlled. Inside the box are molecular machines, bound raw material molecules in transit, and a vacuum. The environment for molecular manufacturing is neither air nor water, but a high vacuum.

The kinds of molecular machinery in the box are analogous to the kinds of machinery that one finds in a macroscopic factory. For example, conveyor belts and rollers are an excellent way to transport and process molecules in molecular manufacturing. Molecules can be bound to a belt, transported by motion of the belt, and two belts coming together can force a collision of two reactive molecules in a particular orientation, causing a transformation to occur, with the resulting product exiting on a belt. Different arrangements of belts can be used to arrange for collisions of longer duration and for transforming high-frequency reactant streams into several low-frequency streams, slow-speed streams into high-speed streams, and other variations that are necessary to make a system work in a well-balanced way.

Processes like these can be used to make arbitrarily complex structures. Two belts meet, putting two building blocks together to make a larger building block. Later, other blocks are joined to make a larger block, and then the larger blocks can be pressed together, bonding them to form a still larger block. Therefore, one can design a hierarchical, convergent process in which molecular fragments are assembled to form ever larger building blocks until macroscopic blocks are joined to make a macroscopic product.

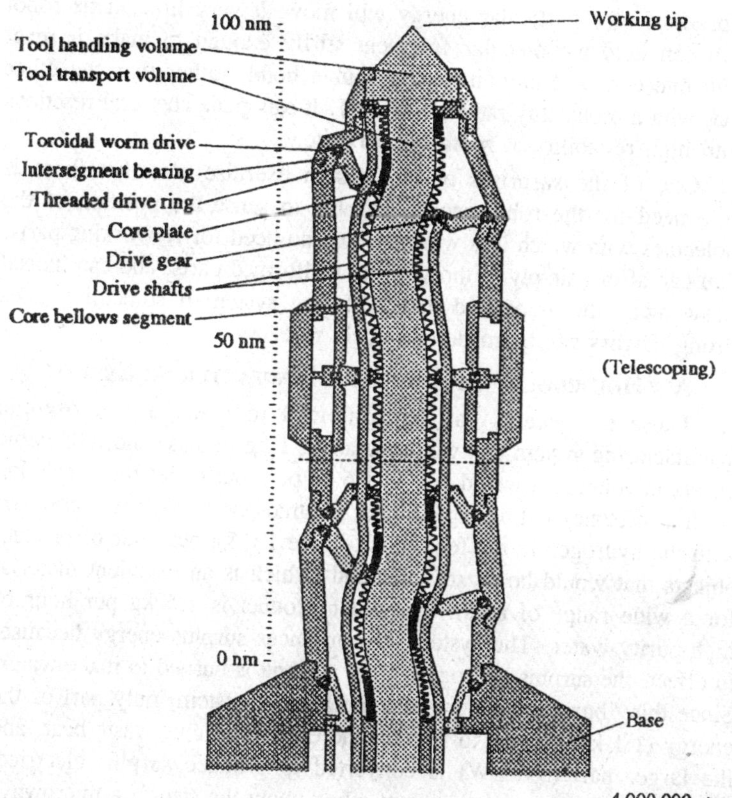

Fig. 1.3. A design for a stiff arm for a nanomechanical robotic manipulator.

A basic problem with such a scheme is that special-purpose machinery is always larger and more complex than the product it makes (unless the product has highly redundant substructures). To make a wide range of products, or to have a simple device make a more complex device, there is a strong incentive to use a device more like a hand—a robotic manipulator. Fig. 1.3 shows a cross section of such a manipulator in two different configurations. The robotic arm pictured has six degrees of freedom and is 100 nanometers long.

The robotic arm is designed to be a very stiff structure to minimize errors caused by thermal vibrations. A soft spring permits large displacements in response to thermal vibrations of a particular energy, giving poor control of the position of a reactive molecule on the end of the spring. If a molecule is held by a stiff spring, however, then vibrations of a particular energy will move it very little. This robot arm can hold a molecular fragment stiffly enough to make it react with one end of a carbon-carbon double bond, rather than the other end, with a probability ratio of 10^{15} to 1. It can guide chemical reactions with high reliability at room temperature.

One of the surprises of this design exercise was that there is little need for the robot arm to be able to sense the position of the molecules with which it is working, and no need for re-working parts. You can afford simply to throw away malformed parts, and can indeed throw away the associated manufacturing system if something goes wrong. Errors can be made extremely rare.

A "Household" Molecular Manufacturing System

These subsystems can be combined to make a macroscopic manufacturing system that would be about 1 kg in mass and 0.05 cubic meters in volume. It would take in 1.6 kg per hour of feedstock solution (such as acetone) and 0.9 kg per hour of atmospheric oxygen to combine with the hydrogen in the feedstock to give 1.0 kg per hour of product objects that would be mostly diamond (which is an excellent material for a wide range of uses). The other product is 1.5 kg per hour of high-purity water. The system also produces surplus energy because, in effect, the surplus hydrogen in the acetone is burned to make water. Since this "burning" occurs in a mechanical system, only part of the energy (1.1 kW) is lost to friction and entropy to give waste heat, and the larger part (3.6 kW) is converted to produce surplus electrical power. The resulting picture is of a box about the size of a microwave oven, with a fan to dissipate waste heat, and plugged into the wall to get rid of surplus electric power.

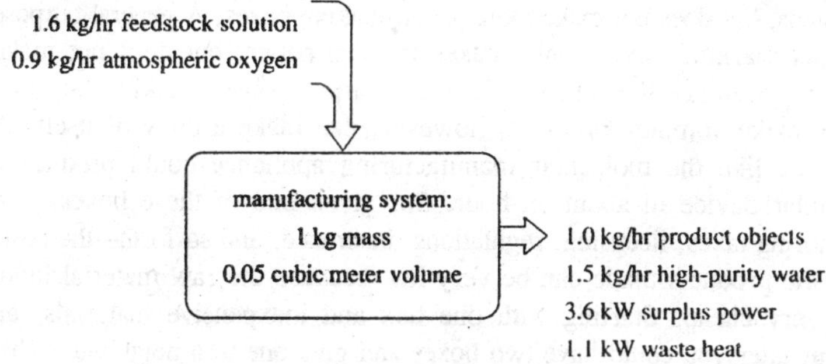

Fig. 1.4. Molecular manufacturing appliance.

A few comments will illustrate the potential significance of this appliance. Once upon a time, if you wanted to have the sound of a flute ill your home, you needed a flute and a flute player. If you wanted to add the sound of a violin, you needed to add a violin and a violin player. If you wanted your home to have the sound of an orchestra, it helped to be a king. Today, after a series of steps in technological evolution, you can buy a box that can read a vast amount of digital data from a CD and use it to control a general-purpose sound-making device. As a result, you can have any sound that you want in your home.

Once upon a time, if you wanted to make patterns of marks on a piece of paper such as a mark the shape of a letter "A," you needed a block of metal cut in the shape of a letter "A." To make a "B," you needed another block of metal in the shape of the letter "B," and so on. To set a page of type took a major capital investment. Today, it is possible to buy a system that stores a large amount of digital data and that drives a general-purpose mechanism for making little black marks on paper (a laser printer, for example). So now we have general-purpose paper-marking devices.

Today, if you want to make metal objects of one shape, you need a lathe. To make objects of another shape, you need a milling machine, and so on. You need a whole host of special-purpose machines for manufacturing the goods of an industrial civilization. What we are now looking toward, however, is a transition to a world in which digital data can be used to control general-purpose machines that will put the fundamental building blocks of matter in place to build almost anything.

There is one large difference between this transition and the two earlier transitions. A general-purpose sound-making box only makes sounds, but does not make more sound-making boxes. A general-purpose paper-marking device only makes marked paper, but does not make more machines like itself. A general-purpose object-making box that can make complex products, however, can make a copy of itself. A device like the molecular manufacturing appliance could produce a similar device in about an hour. So, given one of these boxes—and ignoring taxes, licensing, regulations, insurance, and so forth—the costs of the products made can be very low because the raw material input is very cheap. Starting with one box and inexpensive materials, an hour later you could have two boxes and give one to a neighbour. This has enormous economic consequences.

Characteristics of Molecular Manufacturing

The advantages of molecular manufacturing extend to products over a wide range of sizes, from nanoscale instruments (well-matched to the challenging and complex problem of repairing molecules in living cells), up through the kilo gram scale. These systems can be extended, with comparable efficiency, to products larger than jumbo jets.

Table 1.1. Characteristics of molecular manufacturing

Characteristic	*Result*
Strong materials	About 80 times the strength of aluminum
High component densities	About $10^{26}/m^3$
Near-perfect quality	Nanoscale defects
Cheap inputs	Bulk industrial chemicals
Fast	From a finished design to manufactured products in an hour
Productive	Processing its own mass in an hour
Low cost	Tens of cents per kilogram
Wide size range	Nanoscale to jumbo jet

For perspective, molecular manufacturing can be compared with more familiar systems. Conventional engineering systems (such as aircraft, computers, factories, and so forth) share several characteristics. These include being composed of many distinct parts, held together by rigid supporting structures. The parts (or, in computers, electric currents) move along controlled paths with only one or two degrees of freedom. On the other hand, if you look at the molecular world today (at chemistry and biology), you often find molecules tumbling freely in solution. If macroscopic engineering worked this way, you would open up the hood of your car, and inside you would see a seething jumble of parts.

In materials science today, you find molecules locked in solid, rigid structures. If this were the way macroscopic engineering worked, you would lift the hood of your car and find a solid, immobile block of steel. Although amazing things are being done in chemistry and biology (tumbling molecules can be made to go together to make useful things), and although materials science is the foundation of our macroscopic machine technology and of our electronics technology, both of these fields are quite unlike conventional engineering.

Nanosystems shares many characteristics of conventional engineering systems. They will be used to make instruments, computers, factories, and so forth. Like conventional engineering systems, they will have many distinct parts, rigid supporting structures, and controlled motions and (in electronic systems) controlled electronic activity. Thus, in their architectures, nanosystems will be conventional engineering systems. The proposal, then, is to make systems much like those we are familiar with on the macroscopic scale—factories that have parts (and some way of picking up the parts and putting them where you want them), or devices that have a rigid framework and controlled motions (so that you can design them and have them work, but on a far smaller scale), but designed using the basic building blocks of matter.

Computer Revolution

We can make a few comparisons of molecular manufacturing with the computer revolution. Before the revolution, there were a series of special-purpose machines, adding machines, analog computers, punch-card tabulators, and so forth. Afterward, these were replaced by general-purpose information-processing machines based on small, fast devices that handled information in basic units (bits and words). With molecular manufacturing, special-purpose machines (such as lathes, injection-molding machines, and stamping machines) will be replaced by general-purpose material-processing machines based on small, fast devices that handle matter in its basic units (atoms and molecules).

There are strong parallels with the computer revolution not only in speed, in generality, and in using basic units, but also in the precision of digital systems (working in a domain where results are either exactly right or exactly wrong) and where good design practice can make the probability of being exactly right very high at each step.

Current Technology

What is missing on the path to molecular manufacturing? Today, we still do not have molecular manipulators. It seems that these can be built in the near term, and can bring substantial short-term commercial benefit, providing a new generation of instruments. A step beyond initial devices will be the use of molecular manipulators to assemble building blocks to make complex structures. Building blocks are presently being designed by Markus Krummenacker of the Institute for Molecular Manufacturing. More design work is needed, followed by experimental synthesis and use.

Public Policy

Given the large-scale implications of molecular manufacturing, there is a need for analysis of policy options for the development, regulation, and internationalization of these technologies. There is a real danger of hostile competition developing when a new technology emerges with military applications. There is a need for understanding of these issues among policy-makers and the public. This conference is a part of that process. There is also a need for organizational structure and management for the next stages. That means more money and people.

Conclusions

Many of the consequences of molecular manufacturing can be summed up as follows.

1. *A replacement for industry as we know it*. As a new way of making products of unprecedented performance, quality, and low cost, molecular manufacturing appears to be a replacement for industry as we know it.
2. *A sustainable basis for global wealth*. Because the raw materials for molecular manufacturing can be inexpensive and common (and can even include carbon dioxide), because the products can include low-cost solar cells with mechanical properties that permit you to use them as a surfacing material on roads, and because of the efficiency of these technologies, molecular manufacturing can provide a sustainable basis for global wealth.
3. *An opportunity to roll back environmental impact*. Because of those same characteristics, molecular manufacturing offers an opportunity to roll back environmental impact. This is why members of the environmental community have become increasingly interested in this technology not as a solution to environmental problems, but as a help to those who are trying to solve them. By making it easier for people to get what they want with less environmental impact, a given amount of political pressure for a clean environment will tend to go further.

4. *An enormous challenge to security*. Anything that can make extremely high-performance hardware rapidly, at very low cost and in large quantities, presents an enormous challenge to security in a military sense. It could lead to an unstable arms race.
5. *An enormous challenge to our institutions*. Anything that involves the changes suggested by what has been enumerated here presents an enormous challenge to our institutions.
6. *A point of unprecedented global leverage*. Right now, the level of activity in molecular manufacturing and in consideration of its consequences, is small. Yet, in the longer term, the consequences will be enormous. Therefore, it is believed that activity in this area, advancing these developments, and helping to guide them in a responsible fashion, offers to all of us a point of unprecedented leverage over global problems.

2

Ocean of Molecules

Protoplasm is a highly complex mixture of some elements and compounds found in the bodies of living beings. The protoplasm is variously known as the *living matter*, *living substance* or *physical basis of life*. It is the basic fundamental substance exhibiting all the vital processes of the cell. The protoplasm was first observed by *Corti* in 1772. In 1835 *Dujardin* a Frenchman described it as a soft and gumy substance and named it as *scarcode* i.e. *flesh*. *Purkinje*, a Bohemian physiologist (1839), was the first biologist, who gave the name *protoplasm* to this living substance. *Hugo von Mohl,* a German botanist, in 1839, also suggested the name *protoplasm* for the granular and viscous substance found in plants similar to that found in the animals. The above-mentioned German botanist popularized the word protoplasm as a name given to the living matter found in plants and animals. Protoplasm is the most complex and interesting substance. It is not to be thought of a chemical compound but rather as very complex organized system. Protoplasm varies somewhat in its nature from cell to cell and from organism to organism, but basically it must be the same, as evidenced by its common manifestations of metabolism, growth, reproduction and by some other peculiarities.

Physical Nature of Protoplasm

Different workers have proposed different theories to explain the physical nature of protoplasm as given under the following heads:

Granular Theory

This theory was propounded by *Altmann* in 1893. According to this theory, protoplasm consists of numerous tiny granules as shown in *Amoeba*. *Henle, Maggi,* etc., considered these proto-plasmic granules

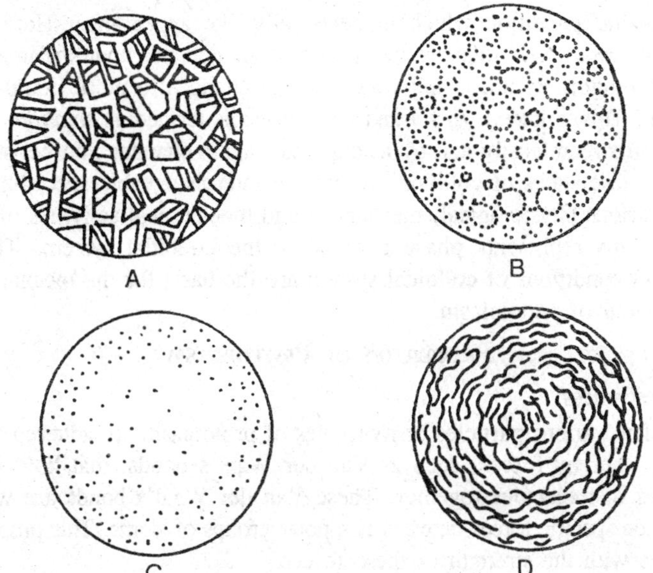

Fig. 2.1. Physical appearance of protoplasm. A—Reticular, B—Alveolar, C—Granular, D—Fibrillar.

plastidules. Altmann reconginzed them as "elementary organisms," or *bioplasts* (or cytoplasts).

Alveolar Theory

The alveolar nature of protoplasm was suggested by *Butschilli* in 1892. According to him, protoplasm consists of many suspended droplets or alveoli or minute bubbles, resembling to foam of emulsion.

Fibrillar Theory

This theory was put forward by *Flemming*. According to him, protoplasm consists of fibres embedded in the inner mass of matrix. The fibrillae are called mitome or spongioplasm and ground substance is termed paramitome or hyaloplasm.

Reticulate Theory

This theory was proposed by *Klein, Cornoy* etc., suggesting that the protoplasm consists of a network or reticulum of fibres in the ground substance.

Colloidal Theory

Proposed by *Wilson* in 1925. According to this theory the protoplasm having any of the aforesaid appearances is always a fluid-colloidal system having various chemical inclusions in *gel phase*. A gel is a semi-solid condition which presents jelly-like appearance. Here the molecules are held together by various bonds depends upon the imature and strength. By absorbing water the gel changes to more liquid-like phase. This is known as *sol* and the process is termed as *solation*. The sol can stream easily and by losing water again changes into gel-state. Both these states of colloidal matrix are interchangeable according to the various physiological, mechanical and biochemical activities of the cell. This represents phase reversal in the colloidal system. These sol-gel conditions of colloidal system are the basis for the mechanical behaviour of protolplasm.

PROPERTIES OF PROTOPLASM

Cohesiveness

The various particles or molecules of protoplasm are adhered with each other by forces, such as Van der Waal's bonds, that hold long chains of molecules together. These Van der Waal's bonds are weak and non-specific forces between non-polar groups of atoms. This property varies with the strength of these forces.

Contractility

This property is significant in various stomatal operations in plants. The contractility of protoplasm is important for the absorption and removal of water as they generally occur in protoplasm.

Electrical Charge

Protein molecules repel each other because their overall charge is similar. All molecules are either positively charged or negatively charged. However, if the molecules approach one another close enough so that valency forces can act, then they may be attracted to each other.

Precipitation

Addition of certain amount of electrolyte in a colloidal suspension causes its dispersed particles to colloide, aggregate and finally to precipitate as suspension. For example, the addition of HCl to a colloidal system of arsenic sulphide causes precipitation.

Viscosity

The viscosity of the ground substance of the cell varies greatly. It may be as low as that of water, or may be very high in the gelating cytoplasm of pseudopodia of *Amoeba*.

Streaming Movement or Cyclosis

The protoplasm exhibits various sorts of streaming movements inside the cell. These have been studied in *Amoeba, Paramecium* etc. The movement involves only the localized portions with no visible changes in the protoplasm. No complete explanation of this has been given as yet. But it is seen that its rate depends upon the rate of cell metabolism. It is due to the fact that the energy is supplied by respiration.

Amoeboid Movement

The amoeboid movement as exhibited by *Amoeba* and other protozoans involves the movement of entire protoplasm of the cell, where the cytoplasm moves as one mass carrying the various inclusion with it. It is due to the continuous change of gel to sol and sol to gel.

Brownian Movement

It is characterized by the zigzag motion of suspended colloidal particles, occurring due to the bombardment of one particle or molecule by other. This type of movement of particles was first of all observed by *Robert Brown* in 1827 in the colloidal solution and hence such movements are known as Brownian movements. The higher the temperature, more rapid the movement and thus viscosity of cell is decreased. This means that high viscosity indicates a more gel-like state of protoplasm and low viscosity, a more sol-like condition.

Tyndall Effect

Colloidal particles of protoplasm have the property of scattering light. When a beam of light is passed through a colloidal solution it becomes visible. This is the Tyndall effect. A colloidal solution of proteins in water shows a typical Tyndall cone.

Adsorption

The tendency of particles, molecules or ions to adhere to the surface of certain solids or liquids is known as adsorption and is exhibited by the particles of colloidal system. The phenomenon helps the matrix to form protein boundaries.

Biological Properties

Protoplasm has all the biological properties of a living organism. It is capable of nutrition, respiration, excertion, metabolism, growth and reproduction. It has the property of irritability, e.g., it responds to stimuli like heat, light and chemicals. It also has the property of conductivity, i.e., of conducting impulses produced by stimuli.

Chemical Nature of Protoplasm

For the detailed study the chemical components of the cell can be classified as inorganic (mineral salts) and organic (proteins, carbohydrates, nucleic acids, lipids and so forth) substance. Although the most prominent constitutent of protoplasm is water—the substance which gives protoplasm its characteristic structure is protein. Lipids are important in all membranes and carbohydrates serve as nutrient stores. The protoplasm of a plant or animal cell contains 75 to 85% water, 10-20% protein, 2-3% lipids, 1% carbohydrates and 1% inorganic material. The following Table 2.1. gives approximate active protoplasm.

Table 2.1. Chemical Analysis of Protoplasm

Substance	*Percent*	*Average molecules weight*	*Number of molecues in relation to protein*
Water	85	18	180
Protein	10	36000	1
DNA	0.4	10	—
RNA	0.7	4.0×10^4	—
Lipid	2	700	10
Other Organic matter	0.4	250	20
Inorganic substances	1.5	55	100

Water

Water is a substance that is justifiably called the *fluid of life*. It comprises over 90% of the chemical content of most organisms. In human body about 55% of the water (20-22 litres) is intracellular water and remains confined to the cells and the rest is found in extracellular fluids such as blood, tissue fluid and lymph. Water participates directly or indirectly in all metabolic reactions. The biologically active conformations of macromolecules and arrangement of phospholipids in the lipid bilayer of membranes are dependent on water. Water helps to keep minerals ionised in body fluids. It ionises itself to provide hydrogen ion (H^+) concentration to body fluids. It also helps in maintaining the constancy of the internal environment of an organism.

Water is a remarkable compound with unique properties that result from its molecular configuration and hydrogen bonding. Some important properties of water are as follows.

Polarity

The water molecule is composed of two hydrogen atoms covalently bonded to one side of an oxygen atom. Since the mean angle (105°) between the hydrogen atoms is not rigid, water can absorb large quantities of heat and be subject to other physical stresses without breakage of the bonds. Water is a polar molecule and as with other polar molecules it has a surface charge. Water is a dipolar substance in that the hydrogen pole is positively charged and the other pole is negatively charged due to electrophilic (electron attracting) properties of oxygen. The polar nature of the water molecule causes salts to be held in solution through charge interaction (ionization; salt dissolved in water exist in the form of positive and negative charges).

Cohesion and adhesion

Because of the asymmetrical distribution water molecules associate with each other (cohesion) and 'wet' other substances (adhesion). The attraction of the positive hydrogen atom of one water molecule for the negative oxygen atom of another water molecule results in a *hydrogen bond.* Hydrogen bonding of water produces a dipolar molecule and favours the formation of a lattice-like structure that enables the packing of many atoms into a small area and stabilizes the molecular structure of water. Fluidity of water is maintained by very rapid formation and dissociation of hydrogen bonds between water molecules.

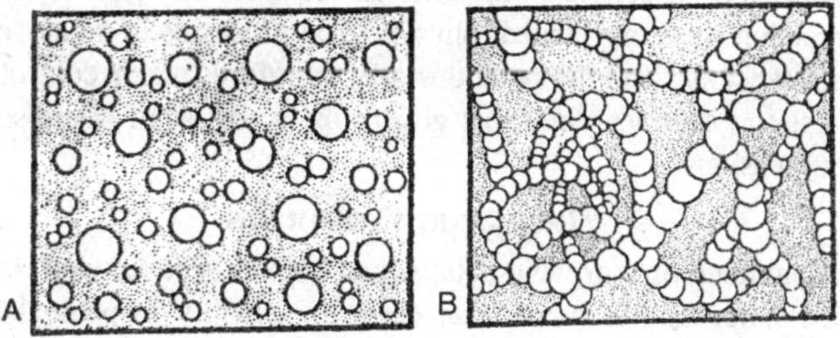

Fig. 2.2. Two water molecules showing polarity and formation of a hydrogen bond between them.

Latent heat and specific heat

The presence of hydrogen bonds is directly responsible for the high heat of fusion, high specific heat, and high heat of vaporization of water. The energy required to break hydrogen bonds for melting of ice, heating of water, and evaporation of water is considerably greater than the energy needed to overcome the Van der Waal's forces that are normally found in the weak association of molecules of ethane, ether and benzene. Due to high specific heat, water affords protection against sudden temperature changes in living organisms. High latent heat of evaporation of water causes elimination of excess heat through evaporation of sweat. This maintains a constant body temperature.

Density of freezing properties

The density of water decreases below 4°C and ice therefore tends to float. It is the only substance whose solid form is less dense than its liquid form. The fact that water below 4°C tends to rise helps to maintain circulation in large bodies of water. This may result in nutrient cycling and colonisation of water to greater depths.

Solvent action

A property of water that is very important to the living cell is its solvent action. Because it forms a solution with a vast array of compounds, water is sometimes referred to as the *universal solvent*. The solvent action of water is an effect of its ability to form hydrogen bonds due to the asymmetrical distribution of its charges. In solution with water, compounds like sugars, alcohols, and amino acids—which contain oxygen atoms, hydroxyl (-OH) groups, amino ($-NH_2$) groups—form hydrogen bonds with the molecules of water.

The solvent action of water is of tremendous importance for the living system. The essential elements necessary for normal growth, the compounds necessary for energy transfer and storage, and the components of structural compounds all require water as a translocation and reaction medium. Indeed, physiological processes in dilute solutions and suspensions; and the reactions are, therefore, under control of the physical and chemical laws that govern the activities of dilute solutions and suspensions.

Inorganic Compounds

A number of inorganic salts occur both in free as well as in the ionised state. The elements that occur in quantity in protoplasm are oxygen, carbon, hydrogen and nitrogen. Always present but in much smaller amounts are potassium, phosphorus, calcium, sulphur, magnesium and iron. These ten are generally referred to as *essential elements*. Chlorine and sodium are necessary components of most animal protoplasm but are apparently not essential for plant protoplasm. Copper, boron, iodine, maganese, zinc and several other elements which are present in all protoplasm but only in minute quantities, are called *trace elements*. This term must be used with care, because these elements are still important even though they are present only in very small quantities. The essential elements found in human protoplasm are listed in table 2.2.

The elements are almost always present in protoplasm in the form of chemical compounds rather than elements. Many of these compounds inorganic salts, which are usually in solution. These salts have numerous functions. They serve as source of elements to be built into other compounds, and some act as *buffer* in maintaining the proper acidity or alkalinity in the protoplasm. Because salts in solutions are electrical conductors, they also function in some way in connection with the electrical properties of protoplasm, however, this is not well understood at present.

Table 2.2. Showing Percentage by Weight to the Element in Protoplasm

Elements	*Symbol*	*Percent*
Oxygen	O	65.04
Carbon	C	18.24
Hydrogen	H	10.05
Nitrogen	N	3.15
Potassium	K	1.60
Phosphorus	P	0.84
Calcium	Ca	0.25
Sulphur	S	0.20
Magnesium	Mg	0.04
Iron	Fe	0.01
Chlorine	Cl	0.27
Sodium	Na	0.26

ORGANIC COMPOUNDS

Proteins

Protein is an indispensable constituent of the diet because it is the only source of the amino acids. Amino acids are needed to built up new tissue during he period of growth; to maintain the structure of every tissue cell including its content of protein—containing enzyme systems; to provide raw material for the manufacturing of digestive enzymes and certain hormones; and to maintain the normal concentrations of plasma proteins and haemoglobin.

The proteins are macromolecules of very high molecular weight. The polymer molecules of protein are formed by linking together of a number of amino acids. About 20 odd different kinds of amino acids are known. Each amino acid has an amino (NH_2) and a carboxyl (COOH) group. A bond is formed between the amino group of one amino acid with the carboxyl group of another amino acid. The bond is called peptide bond. With the help of such peptide bonds a variety of amino acids can form a long chain molecule called polypeptide.

The nutritional value of a protein depends upon its amino acid composition. Synthesis of amino acids from organic keto-acids is of common occurrence in bacteria and plants. Animals can synthesize only about half of the naturally occurring amino acids (20). Thus there are two categories of amino acids:

Essential amino acids

The amino acids which cannot synthesized by an organisms i.e. they must be obtained from the dietary proteins are called essential amino acids. These are *arginine*, *histidine*, *isoleucine*, *leucine*, *lysine*, *methconine*, *theronine*, *phenylalanine*, *tryptophan* and *valine*.

Non-essential amino acids

Although these amino acids are required by the animal as they are found in the protein of the tissues, but they can apparently be synthesized from alpha-keto acids, by the process of amination. These amino acids are *tyrosine*, *serine*, *alanine*, *asparagine*, *proline*, *hydroxyproline*, *asparatic acid*, *glycine*, *glautamine*, *glutamic acid* and *cystine*.

People who eat diets lacking only one of the essential amino acids are unable to synthesize normal body proteins and become protein deficient, despite having a normal nitrogen intake.

Most animal proteins, are *complete or first class proteins;* that is they contain all the necessary amino acids in appropriate proportions for ultilization by man as eggs, meat, kidney, fish, liver, poultry, milk etc.

Plant proteins are *incomplete* or *second class proteins,* since they lack certain essential amino acids. as. cereals, nuts, legumes etc. Thus people living on a strict vegetarian diet may become protein deficient.

Associated with protein deficiency is *Kwashiorker,* which affects millions of children in developing countries. It must commonly occurs in infants after weaning and particularly when given an inadequate and predominantly carbohydrate diet. Many of the children also have an inadequate energy intake and the disorder is known as *protein-energy malnutrition* (PEM). The affected child stops growing and loses weight. The skin and hair may become depigmented, and oedema associated with low plasma levels of albumin often develop. Untreated Kwashiorker is often fatal.

Structure of Protein

The critical determination of biological function of a protein is its conformation, which is defined as the three-dimensional arrangement of the atoms of a molecule. Four basic structural levels are assigned to proteins. These are:

Primary Structure

Arrangement of amino acid in a polypeptids chain is the primary structure of a protein. Amino acid sequence of the smallest protein known, that is of insulin, was studied by *F. Sanger* (1954). Insulin consists of two chains, A and B, joined together by two disulplide bonds (A chain contains 21 and B chain 30 amino acids). Normal human haemoglobin consists of four polypeptide chains (2α and 2β) to be held together by non-covalent forces. Each α-chain is composed of 141 amino acids, and each β chains 146 amino acids.

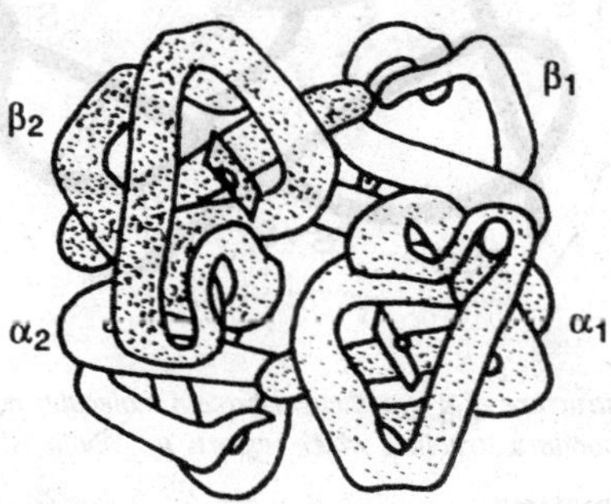

Fig. 2.3. Structure of haemoglobin; the molecule consists of four chains: two α-chains and two β-chains.

Secondary Structure

The polypeptides are held together to give rise a definite shape (helical pattern) of the protein molecules by hydrogen bonds. Hydrogen bonds can be broken by many physical and chemical treatments, for example, by excessive heat, pressure, *pH*, electricity, heavy metals and other agents that changes the environment of protein.

The most common type of secondary structure is α-helix. This is formed by the bending of polypeptide chain to form hydrogen bonds in the same chain. The bending of the chain is very regular and chain assumes the shape of a helix.

Pauling and *Corey* have considered an alternate type of secondary structure. In this case the polypeptide chains are lead linearly either parallel or antiparallel with respect to one another. This is called β-structure of β-helix.

Tertiary Structure

The arrangement and inter-relationship of the twisted chains of protein into specific loops and bends in called the tertiary structure. Such a structure enables the proteins to form specific layers, crystals or fibres. The tertiary structures is maintained by hydrogen bonds disulphide bonds, Van der Waals interaction, hydrophobic interactions and ionic bonds. The tertiary structure is important and found in globular proteins. If this structure is disrupted, the biological activity of protein would be lost.

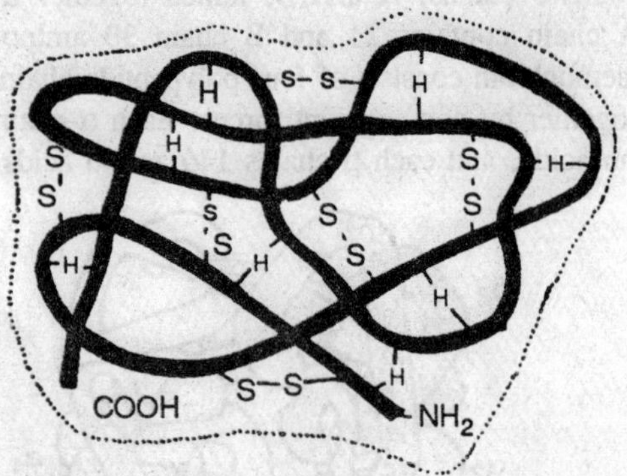

Fig. 2.4. Tertiary structure of a hypothetical protein molecule; note the extensive folding of the secondary structure which imparts a globular shape to the protein.

Quaternary Structure

This defines the degree of polymerisation of a protein unit. This structure is found in only oligomeric proteins. They are composed of

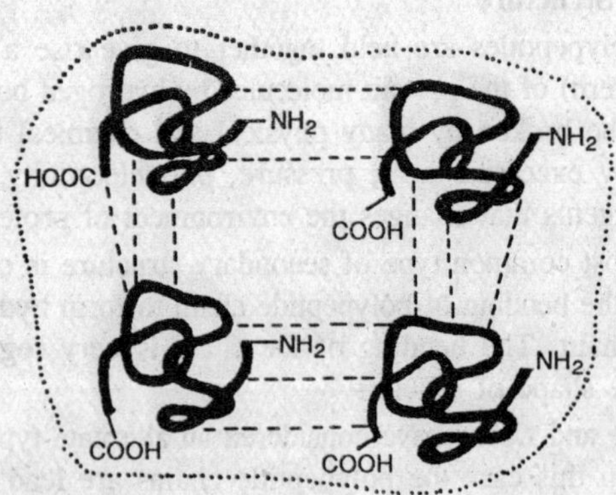

Fig. 2.5. Quaternary level of protein structure; note a diagrammatic representation showing four sub-units which are polymerized in the functional protein molecule.

subunit peptide chains linked together by any or all of the forces that can act between amino acid side chains. Haemoglobin is a fine example of quaternary structure and was studied by *Kendrew* and *Perutz* (1963).

Classification of Protein

The proteins are mainly classified into two—simple and conjugated proteins.

Simple Proteins

These are consisting mainly of peptide chains. On hydrolyses they yield only amino acids. These are of two types:

(A) Fibrous or insoluble proteins.

(B) Globular or soluble proteins.

Fibrous (insoluble) proteins

Fibrous proteins (scleroproteins) comprise all insoluble proteins which have a supporting or protective function in the animals. Secondary structure is the final structure of these proteins. The fibrous proteins can be subdivided into following types:

1. *Collagens.* Approximately 30% of the total protein in mammalian body is collagen. They are the principle proteins of skin, tendon and bones and are resistant to peptic and tryptic digestion. They contain large amounts of hydroxy-proline. They can be converted into soluble gelatin by boiling with water, dilute acids or alkalies.
2. *Elastins.* They are present in ligament, arteries and other elastic tissue. They cannot be converted into gelatin.
3. *Keratins.* They are present in animal skin, nails, hairs, horns, hooves, claws, feathers etc. they are derived from ectoderm.
4. *Myosin.* The major protein of muscles is also a fibrous protein.

Globular or Soluble proteins

These are soluble simple proteins with more or less definite molecular weight. These are speherical or ovoid in shape. They are generally soluble in water. This group includes enzymes, oxygen carrying proteins (globulins), hormones like oxytocin, vassopressin, insulin, glucogon etc.

1. *Albumins.* They are soluble in water, dilute acids and bases. They coagulate when heated. Common examples are albumins of egg white, blood serum and lactalbumin.
2. *Globulins.* They are insoluble in water but soluble in dilute salt solution. They are also coagulated when heated. Examples are serum globulin, fibrinogen, ovoglobulin in egg-yolk etc.

3. *Glutelins*. These are insoluble in water but soluble in dilute acids or base solution. They are commonly found in various plant seeds. Examples are wheat, cereals etc.
4. *Protamins*. These are basic proteins. They are very soluble, small, stable proteins and cannot be coagulated by heating. They are exceptionally rich in arginine amino acid. They are found in the ripe sperm cells of certain fishes.
5. *Histones*. These are water-soluble basic proteins. They coagulate on heating. These are found associated with nucleic acids in nucleoproteins. They are obtainable from the thymus, spleen and nucleated erythrocytes of birds.

Conjugated Proteins

The conjugated proteins differ from the simple proteins in that they consist of proteins combined with some non-protein substances. The term *prosthetic group* is generally used to designate such substances (non-proteinous). They are classified according to the nature of prosthetic group as follows:

1. *Nucleoprotins*. These are formed by the combination of nucleic acids with protein.
2. *Mucoproteins*. They contain carbohydrate bound to protein molecules. The protein components of *mucoids* are combined with large amount (more than 4%) of carbohydrate. Examples are ovomucoid-β from egg white; or osomucoid from blood serum.
3. *Glycoproteins*. The carbohydrate percentage is less than 4% in glycoproteins. The carbonhdrate in these is usually the mucopolysaccharids. Mucin of saliva, chorionic gonadotropins and some other hormones of pituitary such as follicle stimulating hormone and luteinizing hormones are glycoproteins.
4. *Lipoproteins*. They are compounds of lipid and protein. The protein part is water soluble and lipid unsoluble. Examples are cholesterol, phosphatidylglycerides, lipovitelline of egg-yolk and serum lipoporteins.
5. *Chromoproteins*. These are the proteins which have a matelloprophyrin or some similar substance. Examples are haemoglobin which contains a basic protein globin with an iron porphyrin as Heme. Haemocyanin, haemoerythrin, erythrocuorins are other examples.
6. *Phosphoproteins*. In these the protein molecule is linked to phosphoric acid when treated with dilute sodium hydroxide yield inorganic

phosphate. Examples are casein of milk, vitellin of egg-yolk and pepsin.

7. *Metalloproteins*. These proteins contain metals as an inherent part of their molecules. They include tyrosinase (Cu), carbonic anhydrase (Zn), arginase (Mn or Mg) etc.
8. *Flavoproteins*. The prosthetic group of flavin component remains permanently attached to the protein. Flavoproteins act as coenzymes which catalyze oxidation-reduction reactions.

Derived Proteins

These include the products obtained from proteins by the action of heat and other physical agents or by hydrolysis.

Denatured and Coagulated proteins

Upon treatment with certain chemical agents or heat treatment the proteins become insoluble on account of some sort of intramolecular rearrangement and it is said to have denatured and when it gets separated in the form of precipitate, it is said to have undergone coagulation.

Peptides

These include the various fragments of the proteins which are broken off during hydrolysis of the big protein molecules either the presence of the acids or enzymes. The larger fragments had previously been named as the proteoses and the smaller fragments as peptones. This terminology is being abandoned now.

Properties of Proteins

1. ***Colour and Taste.*** Proteins are colourless and usually tasteless. These are homogenous and crystalline.
2. ***Molecular weight.*** The proteins generally have large molecular weights ranging between 5×10^3 and 1×10^6. It might be noted that the values of molecular weights of many proteins lie close to or multiples of 35,000 and 70,000.
3. ***Colloidal nature.*** Because of their giant size the proteins exhibit many colloial propertis such as (a) their diffusion rates are extremely slow, (b) they may produce considerable light— scattering in solution, thus resulting in visible turbidity (Tyndall effect).
4. ***Denaturation.*** In refers to the changes in the properties of a protein. The process of denaturation is followed by coagulation. Denaturation may be brought by a variety of agents. The physical agents, include mechanical action, heat treatment, cooling and freezing operations, ultraviolet rays etc. The chemical agents include X-rays, acetone, alcohol (solvents), salicyclates etc.

5. ***Amphoteric nature.*** The proteins are amphoteric i.e., they act as acids and alkalies both. These migrates in an electric field and the direction of migration depends upon the net charge possessed by the molecule. The net charge is influenced by the *pH* value. Each portion has a fixed value of isoelectric point (pl) at which it will move in an electric field.
6. ***Solubility.*** It is influenced by *pH* solubility is lowest at isoelectric point and increases with increasing acidity or alkalinity.
7. ***Hydrolysis.*** The proteins are hydrolyzed by a variety of hydrolytic agents.

(a) *By acidic agents.* Proteins, upon hydrolysis which come HCl (6-12 N) at 100-110°C for 6-20 hrs yields amino acids in the form of their hydrochlorides.

(b) *By alkaline agents.* Proteins may be hydrolyzed with 2N NaOH. This process is less used as it is highly disadvantageous.

(c) *By proteolytic enzymes.* Under relatively mild conditions of temperature and acidity, certain proteolytc enzymes such as pepsin and trypsin hydrolyze the proteins. Enzyme hydrolysis is used for the tryptophan.

8. ***Reactions involving—COOH group***

(a) *Reaction with alkalies* (salt formation). The carboxylic group of amino acids release a H^+ ion and forms carboxylae (COO^-) ions. These may be neutralized by cations like Na^+ and Ca^{++} to form salts.

(b) *Reaction with alcohols (esterification).* With alcohols, corresponding esters are produced.

$$\text{R—}\underset{\underset{\text{NH}_2}{|}}{\overset{\overset{\text{H}}{|}}{\text{C}}}\text{—COOH} + \text{HOC}_2\text{H}_5 \xrightarrow[\text{catalyst}]{\text{Acid}} \text{R—}\underset{\underset{\text{NH}_2}{|}}{\overset{\overset{\text{H}}{|}}{\text{C}}}\text{—COOC}_2\text{H}_5 + \text{H}_2\text{O}$$

Ethyl ester of AA.

(c) *Reaction with amines.* Forms amides.

$$\text{R—}\underset{\underset{\text{NH}_2}{|}}{\text{CH}}\text{—COOH} + \text{HHN - R} \rightarrow \text{R—}\underset{\underset{\text{NH}_2}{|}}{\text{CH}}\text{—CO—NH—R} + \text{H}_2\text{O}$$

9. ***Reaction involving NH2 groups***

(a) *Reaction with mineral acids* (Salt formation). When AA or proteins are treated with HCl the acid salts are formed.

(b) *Reaction with formaldehyde.* With formaldehyde, he hydroxy methyl derivatives are formed which are insoluble in water and resistant to the attack of microorganisms.

(c) *Reaction with Nitrous acid (Van Slyke reaction).* The amino acids react with HNO_2 to liberate N_2 gas and to produce a-hydroxy acid.

(d) *Reaction with fluro-dinitrobenzene (FDNB)* or *Sanger's reagent.* In mildly alkaline solution, 1-fluro - 2, 4-dinitrobenzene reacts with α-amino acids to produce yellow coloured derivative, DNB-amino acid.

10. ***Reaction involving both COOH and NH_2 groups*** (a) Reaction with triketohydrindene hydrate (*Ninhydrin reaction*). Ninhydrin in a powerful oxidizing agent causes oxidative decarboxylation of α-amino acids producing CO_2, NH_3 and an aldehyde. The reduced ninhydrin then reacts with the liberated NH_3 forming blue-coloured Ruheman's complex.

11. ***Reactions involving—R group or side chain***

(a) *Biuret test.* Compounds containing peptide bonds produce a purple colour when treats with Biuret reagent (.2% alkaline $CuSO_4$). A substance Biuret is formed. Dipeptides do not respond to this reaction.

(b) *Xenthoproteic test.* Yellow colour develops when proteins are boiled with conc. HNO_3 due to presence of benzene ring. This reaction is due to nitration of phenyl rings. This is the test for tyrosine, tryptophan, phenylalanine.

(c) *Million's test.* When proteins are heated with $HgNO_3$ in HNO_3, a red colour develops. This reaction is specific for tyrosine. Tryptophan also responds to this reaction.

(d) *Hoplin's—Cole test* (or *Glyoxylic acid test*). Violet ring develops on addition of Conc. H_2SO_4 at the junction of protein and glyoxylic acid solution. This test is specific for tryptophan.

(e) *Folin's test.* Blue colour develops with phosphomolybdo-tungstic acid in alkaline soln-due to phenol group. Specific test for tyrosine.

Biological Importance of Proteins

Proteins constitute a large part of the cell-structure and are present in all the tissue. Many proteins have special physiological functions.

1. *Membrane proteins.* The integral proteins include translocases, the peripheral proteins include cytochrome—C and monamine oxidase.
2. *Enzymes.* The enzymes are biocatalysts to influence the rats of a chemical reaction. All the enzymes are proteins.

3. *Hormones.* Some of the messangers of our body, the hromones, are proteins.
4. *Blood proteins.* 6 major plasma components are albumen, α-1 globulin, α-2 globulin, β-globulin, γ-globilin and fibrinogen. All are proteins.

CARBOHYDRATES

The class of substances, known as carbohydrates, is comprises of a large number of relatively heterogenous compounds. These are synthesized form carbon dioxide and water in chlorophyll bearing cells during the process of photosynthesis. They are especial constituents of plants (cellulose, starch etc.) but also occur and serve important functions in animals. They serve as the chief source of energy in the food of many animals. Their erergy value is 4 cal per gram and this energy is provided to the various synthetic needs of a cell. In carbohydrate carbon, hydrogen and oxygen are generally found in 1:2:1 ratio.

Structure

Structurally the carbohydrate are the polyhydric alcohols of carbon possessing active aldehyde or ketonic groups, which on hydrolysis yield to such products.

CLASSIFICATION

The carbohydrates may be classified according to their complexity. They are classified into following groups:

1. Monosaccharides
2. Oligosaccharides
3. Polysaccharides.

Monosaccharides

Monosaccharides are those sugars which cannot be hydrolysed into a simple form. Their general formula is $(CH_2O)n$. The value of n ranges from 3-7. If the monosaccharide has an aldehyde group—CHO it is called an aldose e.g. glucose and if a keto group > C = O is present it is called a ketose e.g. fructose.

Monosacchrides include trioses, tetroses, pentoses, hexoses, heptoses. This classification is based on the basis of carbon atoms present in their molecules. The recent trend in the classification of monosaccharides is a combination of both the system. The aldeosugars and ketosesugars are classified as below. The suffix ('- *ore*) denotes that the sugar belongs to the aldosugar grop and the suffix (*-ulose*) denotes the ketosugas.

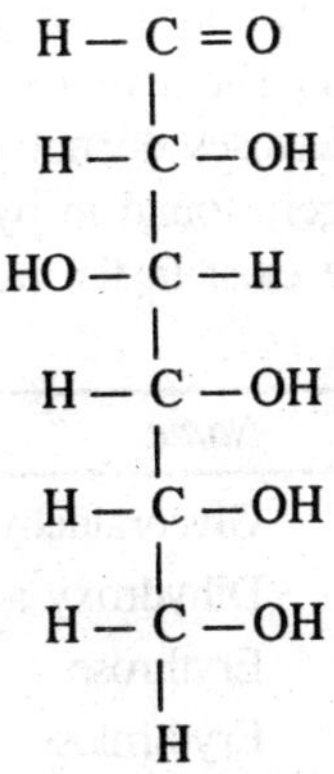

H – C = O
|
H – C – OH
|
HO – C – H
|
H – C – OH
|
H – C – OH
|
H – C – OH
|
H

D-glucose

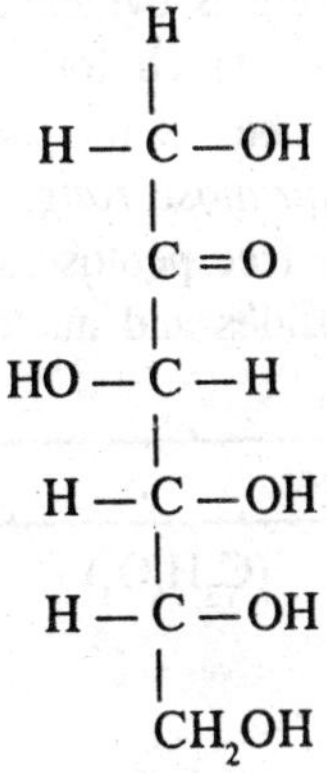

H
|
H – C – OH
|
C = O
|
HO – C – H
|
H – C – OH
|
H – C – OH
|
CH_2OH

D-fructose

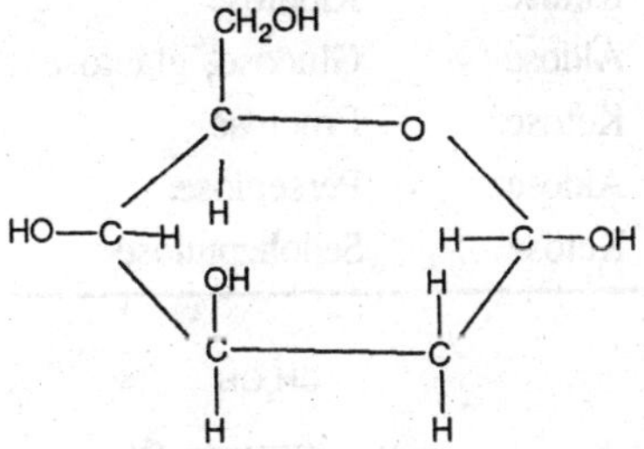

Ring structure of glucose

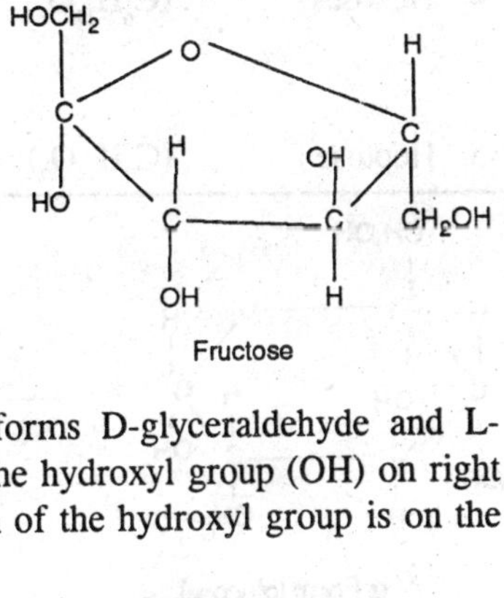

Fructose

Glyceraldehyde occurs in two forms D-glyceraldehyde and L-glyceraldelyde. In D-glyceraldelyde the hydroxyl group (OH) on right side. In L-glyceraldelyde the position of the hydroxyl group is on the left on the asymmetric carbon atom.

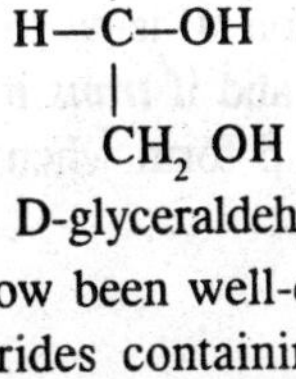

CHO
|
H—C—OH
|
CH_2 OH

D-glyceraldehyde

CHO
|
OH—C—H
|
CH_2 OH

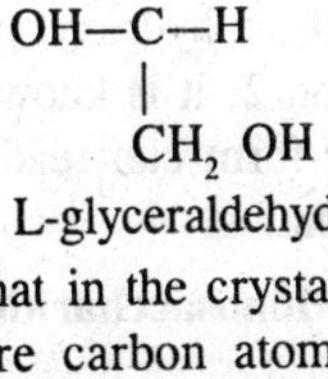

L-glyceraldehyde

It has now been well-established that in the crystalline form, the monosaccharides containing 5 or more carbon atoms exist in the *tautomeric* ring forms. Even in solutions, these sugars mostly occur in the ring forms. The free sugars mostly contain six membered ring in which five members are carbon atoms and the remaining one member is an atom of oxygen. The six membered ring is known as *pyranose ring*. In adition, there is evidence of the existence of five membered rign in the sugar in which four members are carbon atoms and the remaining one member is the atom of oxygen. The five membered rign is known as *furanose rang*. Trioses and tetroses do not possess ring structure. The free pentose sugars are largely found in pyranose forms but in glycosides and nucleic acids these exist in the pyranose form.

Monosaccharides		*Type*	*Name*
1. Triose	($C_3H_6O_3$)	Aldose	Glyceraldehyde
		ketose	Dihydroxy acetone
		Aldose	Erythrose
2. Tetrose	($C_4H_8O_4$)	ketose	Erythrulose
3. Pentose	($C_5H_{10}O_2$)	Aldose	Ribose
		Ketose	Ribulose
4. Hexoses	($C_6H_{12}O_6$)	Aldose	Glucose, glactose
		Ketose	Fructose
		Aldose	Persenlose
5. Heptose	($C_7H_{14}O_7$)	Ketose	Sedoheptulose

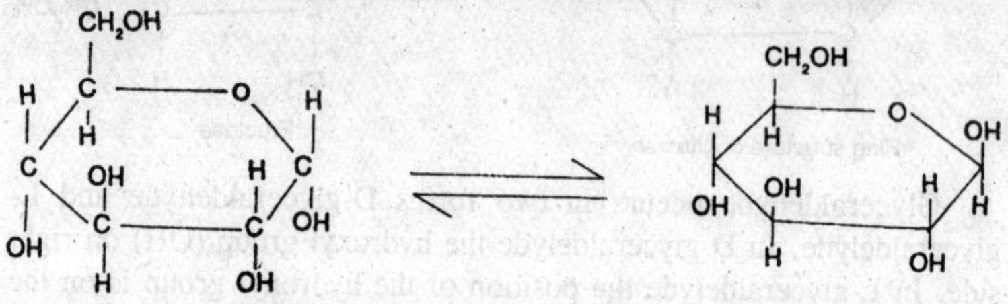

α-Form (glucose) ⇌ β-Form (glucose)

In the pyranose type of hexoses, carbon at 'I' position is asymmetrical. If the hydroxyl group at position 'I' is *cis* to the hydroxyl group at position 2, it is known as α-form and if *trans* it is known as β-form. The α-form can readily pass into β-form when the sugar is brought into solution.

Properties of Monosaccharides

(a) Physical

The monosaccharides are sweet testing, colourless solids. They are soluble in water, partially in alcohols and insoluble in other. When a polarized light is passed through a solution of these carbohydrates, the plane of light is rotated to either right or left side. They contain asymmetric carbon atom hence exist in different isometric forms. The degree of optical rotation may change due to interconversion of isomeres. Fresh solution of glucose gives an optical rotation of +112° which change to + 52.7° on standing. The change in optical rotation is called *mutarotation*.

(b) Chemical

1. *Oxidation.* They can easily be oxidized by oxidizing agents. Glucose yields gluconic acid after oxidation with Tollen's reagent (ammonical Ag_2 O) or Fehlings solution (alkaline $CuSO_4$). The reduction of Tollen's reagent yields silver as polishing on the surface of tube whereas with the Fehling solution red ppt are obtained.

   When strong oxidizing agent is used like conc HNO_3, gluconic acid, is ultimately oxidizes to discarboxylic saccharic acid.

   Glucuronic acid is obtained in animal body on slow oxidation of glucose. Glucuronic acid combines with hormones. It is a major component of hyaluronic acid, heparin, mucoitin sulphate and chondriotine which are found in blood, skin and cartilage.
2. *Reduction.* Free aldehyde and ketone groups of mono-saccharides are reduced to alcoholic hydroxy groups by sodium—mercury amalgam and water. D-glucose after reduction yields a mixture of sorbitol and mannitol.
3. *Condensation.* Aldehyde groups of monosacharides condense with primary amines to form Schiff's base, when treated with hydroxylamine, glucose forms glucose oxime.
4. *Esterification.* The hydroxyl groups of alcohols in the carbohydrate may be converted to esters by treating with the appropriate acetylating agents. When D-glucose is treated with acetic anhydrine in the presence of pyridine, penta-acetyl glucose is formed.
5. *Methylation.* Methylating agents such as Ag_2O, CH_3OH react with monosaccharide to yield glycoside.
6. *Fermentation.* Monosaccharides such as glucose and fructose undergo alcholic fermentation by micro-organisms such as yeasts and produce ethanol and carbon dioxide.

Oligosaccharides

The oligosaccharides are those carbohydrates which on hydrolysis give two to five simple monosaccharide molecules. The oligosaccharides are composed of two to five monosaccharide units. During union of monosaccharde units water molecule is eleminated and the units are linked through an oxygen bridge. It is a glycosidic linkage. Their general formula is $(CH_2O)_{n-1}$.

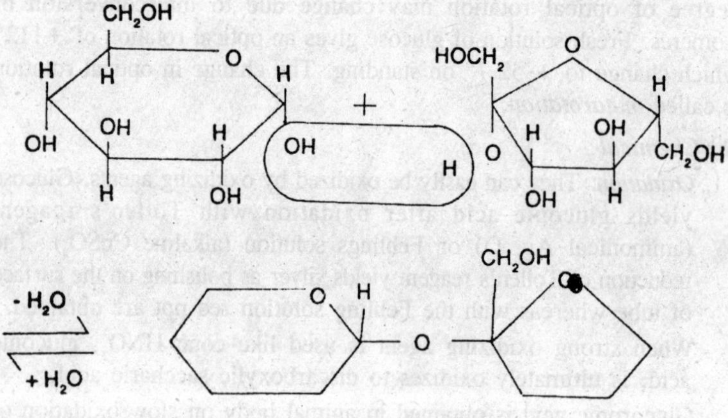

sucrose (diasaccharide)

The disaccharides have been classified into two groups namely, reducing and nonreducing. The reducing disaccharides are maltose, lactose, cellobiose, gentibiose and the non-reducing are sucrose and trehalose.

Cellobiose. Incomplete hydrolysis of cellulose given cellobiose. It is composed of two molecules of glucose linked by β-1, 4 glucosidic bond.

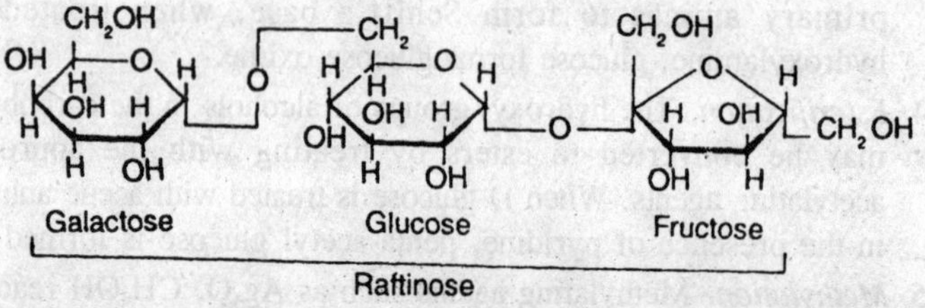

Raftinose

Maltose. It is composed of two units of D-glucose joined together through their 1 and 4 carbon atoms. It is obtained as a hydrolytic product by the action of amylase on strach

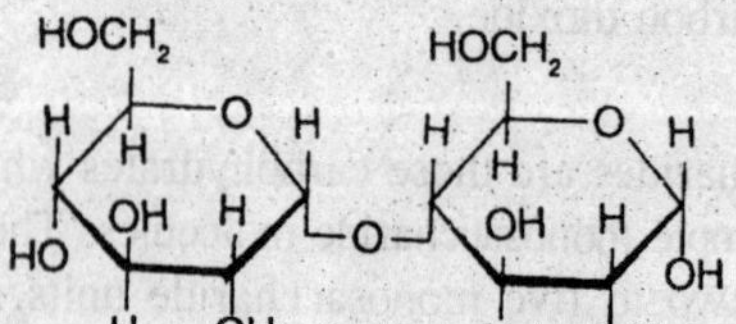

Lactose. Lactose is a disaccharide consisting of glucose and galactose which is synthesized in the mammary glands.

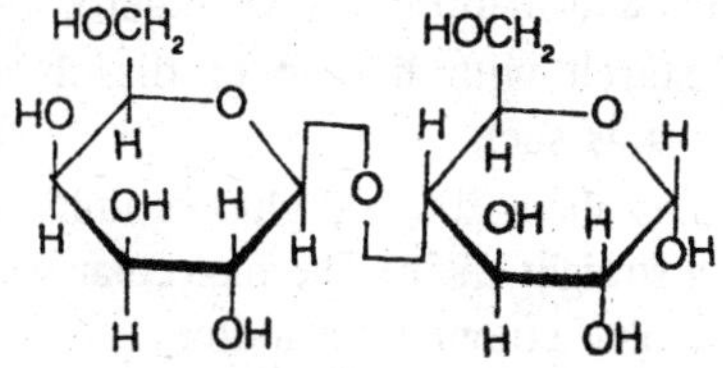

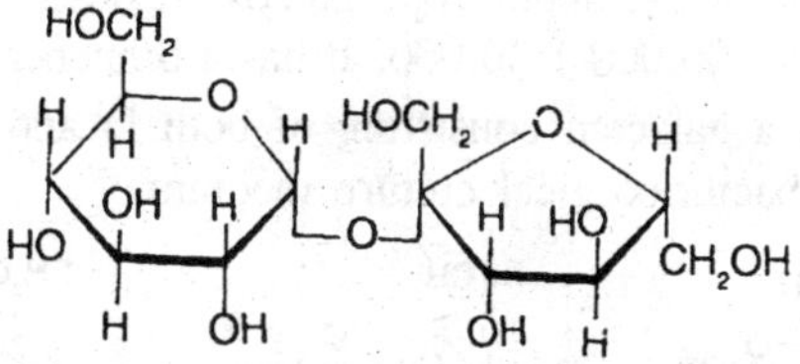

Raffinose. It is a trisaccharide consisting of fructose—glucose—galactose. On hydrolysis it yields *Melibiose* (glucose—galactose) and fructose.

Sucrose. It is formed by the union of one α-D-glucose and one β-D-fructose units with the elemination of one H_2O molecule. Hydrolysis of sucrose by dilute acid or sucrase produces a molecule of glucose and a molecule of fructose.

Polysaccharides

Polysaccharides may be regarded as carbohydrates formed by linking of a number of monosaccharides by glucosidic linkages. A carbohydrate having minimally 6 or more monosaccharide units may be regarded as polysaccharides. Their general formula is $(C_6H_{10}O_5)n$. They are tasteless, colourless amorphous powders which are insoluble in water. because of insolubility and large size, they form colloidal solutions and will not pass across natural animal membranes. They are chemically inert and do not ionize and for this reason very much suited as reserve food material such as glycogen in animals and starch in plants.

Polysaccharides consisting of only one type of monosaccharide units are called *homopolysaccharides* while those with different types of monosaccharde derivatives are *heteropolysaccharides*.

I. Homopolysacchardes

1. *Glycogen.* It forms the carbohydrate reserve of the animal tissues and it is mainly stored in the muscles and liver. Fungi and yeast do also contain glycogen. It is a branched chain polymer having 6000—30,000 glucose units.

2. *Starch.* It is the most important food source of carbohydrate and is found in potatoes, rice, cereals etc. It is hydrolysed in gut yielding dextrins and maltose and eventually glucose.

   Treatment of starch with hot water dissolves amylose, while amylopectine remains as such:

   Amylose. Contains about 200-500 glucose units which are arranged in the form of a straight chain. The molecular wt. is about 150,000. It gives intense blue colour with iodine.

   Amylopectin. It has about 1000 glucose resideus and its molecular wt. is about 200,000-1,00,000. It has a branched structure.

3. *Agar.* It is a galactan consisting of both D and L glactose. It is used as a bacteriological culture medium.

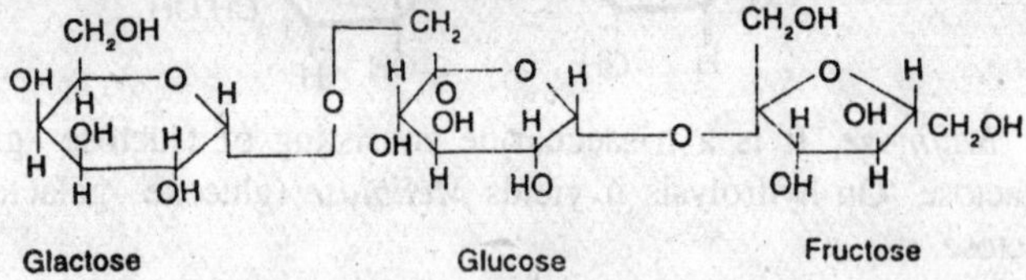

Glactose　Glucose　Fructose

Raffinose

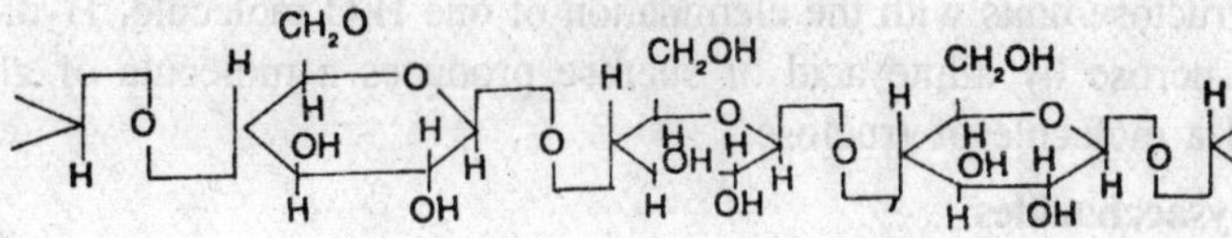

Cellulose

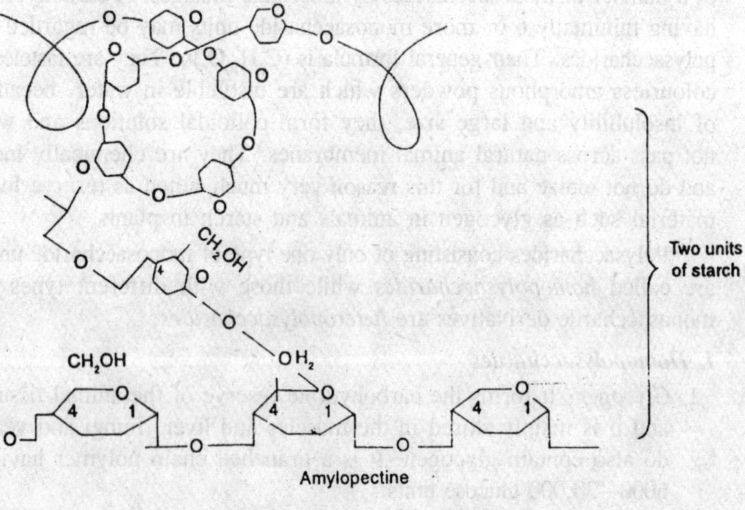

Two units of starch

Amylopectine

4. *Pectins.* These are abundant in fruits, particularly in the rim of citrus fruits like orange and lemons. They contain arabinose, glactose and galacturonic acid.
5. *Xylan.* In addition to cellulose all plants contain xylan. It is a hemicellulose and consists of D-xylose.

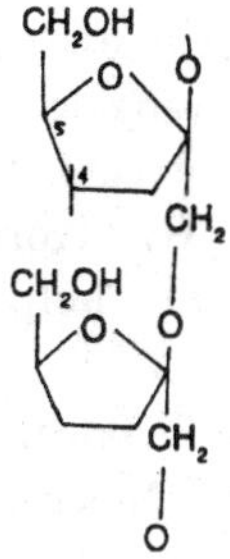

Inulin

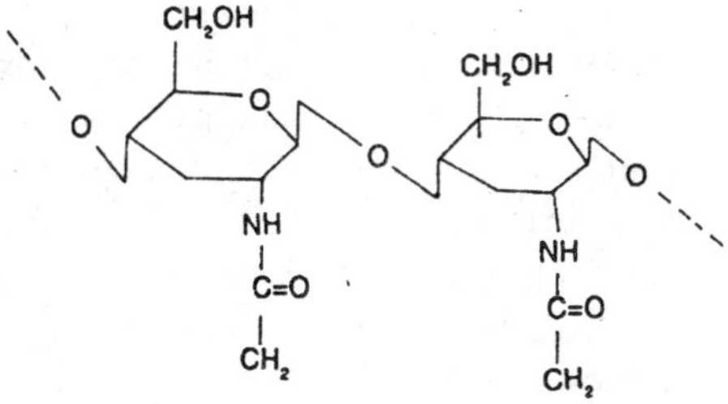

Chitiu

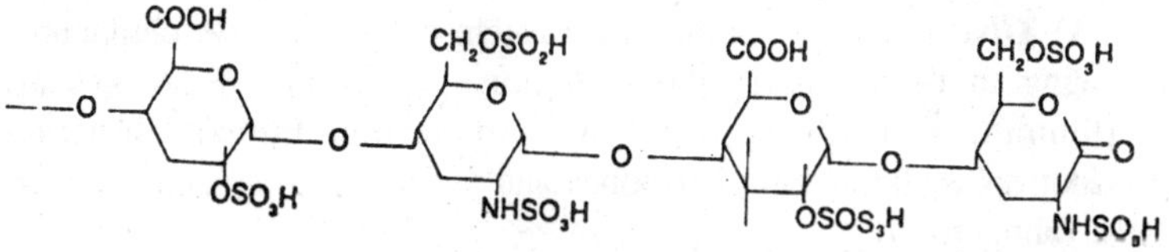

Heparin

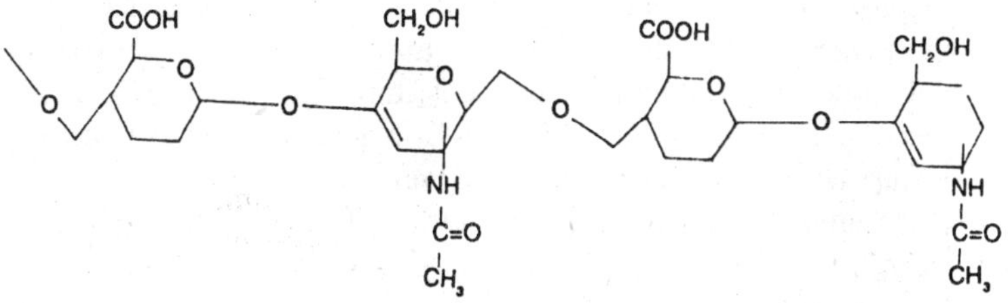

Hyaluronic acid

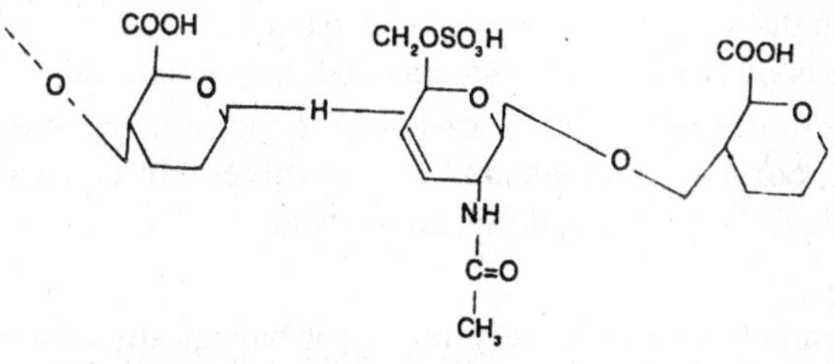

chondriotin sulfate

6. *Inulin.* It is a starch found in the tubers and roots of *Dahlia*, *Dandelions* etc. On hydrolysis it yields fructose. M. Wt. is about 5,000 and about 30-35 fructose units per mole are present.
7. *Dextrins.* They are the intermediate products formed during hydrolysis of starch to the glucose.
8. *Cellulose.* The cell wall of plants is formed by cellulose. The cellulose consists of a lenier chain of glucose units ranging 900-2000. It does not occur in the animal tissue and it cannot be utilized by man for energy production purposes. Herbivorous animals can, however, utilize cellulose with the help of microorganisms as they can digest cellulose.
9. *Chitin.* It is an important structural polysaccharide in invertebrates particularly in arthropods. Structurally, it consists of N-acetyl-D glucosamine units.

II. Heteropolysaccharides

A. ***Glycoproteins.*** These are protein-polysaccharide compounds occuring in tissue, particularly in mucus secretion. Examples are ovalbumin, fibrinogen, γ-globulin of serum, human chorionic gonadotropins, luteinizing hormones and the blood group substances of RBC (antigen).

B. ***Mucopolysaccharides.*** The most important mucopolysaccharides from biological point of view are:

1. *Heparin.* It is present in the liver, lungs and spleen having a molecular wt. about 20,000. The *exect* structure of heparin in unknown. It is found to be constituted by glucuronic acid, glucosamine and sulfuric acid. Sulfuric acid is linked to hydroxyl group of sugar derivatives as well as to amino group of glucosamine. Heparin prevents clotting of blood in the vessels.
2. *Hyaluronic acid.* It is a mucopolysaccharide present in connective tissue and acts as an intercellular connecting material. It is abundently present in the umbilical cord, vitreous fluid of eyes and in the synovial fluid present at the joints. This polysaccharide consists of N-acetyl glucosamine and glucuronic acid.
3. *Chondrotin sulfates.* They are found in the cartilage, adult bones, skin, cornea, tendons and heart valves. It consists of N-acetylgalactosamine and glucuronic acid.

Functions

The carbohydrates are very important biologically as a source of energy for the cell. The main source of energy is glucose. It the main form of crbohydrates which is transported form cell to cell by blood in animal and by sap in plants. Carbohydrates also help in the formation of cell wall in plants. Some carbohydrates may serve as the prosthetic group of certain proteins. Glycogen and starch are major storage materials in animals and plants respectively.

Lipids

Term lipid includes fats and fat-like substances. Strucrurally, the different substances of this group may not be similar to fats, but all are soluble in fat-solvents like alcohols, ether, chloroform, carbon tetrachloride, acetone etc. They are insoluble in water. Lipids, like carbohydrates also, contain carbon, oxygen and hydrogen. The carbon and hydrogen are present in larger quantity than oxygen. Some lipids have phosphorous and nitrogen. The lipids in the cell may serve as condensed reserve of energy as well as form the membranous structures.

Classification of Lipids

Lipids are classified into:

I. Simple Lipids

II. Compound Lipids

III. Derived Lipids

I. Simple Lipids

These are esters of fatty acids with various alcohols.

1. Natural Fats

These are the esters of fatty acids with glycerol. The chemical term for a natural fat is *triglycerides*. The triglycerides are formed by the combination of three molecules of fatty acids joined to one molecule of glycerol. As already stated triglycerides are neurtral fats, neutral because the acid ions are neutralized during their unification with glycerol; fats because the bulk of the molecule is devoid of electro-negative elements which can unite with hydrogen to form water molecule.

$$\begin{array}{lcll} CH_2\ OH & & HOOC{-}R_1 & & CH_2OCOR_1 \\ | & & & & | \\ CHOH & + & HOOC{-}R_2 & \rightarrow & CHOCOR_2 \\ | & & & & | \\ CH_2OH & & HOOC{-}R_3 & & CH_2OCOR_3 \\ \text{Glycerol} & & \text{3 fatty acids} & & \text{Triglyceride} \\ \text{molecule} & & \text{molecules} & & +\ 3H_2O \end{array}$$

Triglycerides can exist in the solid or liquid form. It is based on the kind of fatty acid residues in its structure. The triglycerides which are blow 20°C called *oils* containing a large proportion of short-chained unsaturated fatty acids like *oleic, linoleic acid* etc. The triglycerides which are solid above 20°C are called *fats*. They contain long chained saturated fatty acids like *palmitic acid* and *stearic acid.*

2. Fatty acids

These are obtained by the hydrolysis of fats. These are monocarboxylic acids. The molecules of a fatty acid has a polar carboxyl group soluble in water and a non-polar hydrocarbon chain soluble only in fat solvents.

Table 2.3. Showing a few common fatty acids found in lipids.

Fatty acid		*Formula*
Saturated fatty acid	→	
Butyric acid	→	$CH_3(CH_2)_2$ COOH
Caproic acid	→	$CH_3(CH_2)_4$ COOH
Palmitic acid	→	CH_3 $(CH_2)_{14}$ COOH
Steraric acid	→	CH_3 $(CH_2)_{16}$ COOH
Unsaturated fatty acids		
Palmitoleic acid		CH_3 $(CH_2)_5$ (H=CH $(CH_2)_7$ COOH
Oleic acid		CH_3 $(CH_2)_7$ CH=CH $(CH_2)_7$ COOH
Linoleic acid		CH_3 $(CH_2)_4$ CH=CH CH_2 CH=CH $(CH_2)_7$ COOH
Linolenic acid		CH_3 CH_2 CH=CH CH_2 CH=CH CH_2 CH=CH $(CH_2)_7$ COOH

The fatty acids are classifed under three groups based on their degree of saturation and unsaturation.

(i) *Saturated fatty acids*. A saturated fatty acid contains as many hydrogen atoms as its carbon chain can hold. General formula for saturated fatty acid is R—COOH, where R is CH_3 (CH_2)n. n is varying from zero in acetic acid to 86 in mycolic acid. The most abunant saturated fatty acids in nature *palmitic* (C_{18}) and *Stearic* (C_{16}) *acids.*

(ii) *Unsaturated fatty acids*. They have one or more double bonds in their carbon chain. When there is a single double bond, results in the loss of 2 hydrogen atoms, such fatty acids are called *monous saturated fatty acids*. Their general formula is CnH_{2n}—1—COOH, examples are *palmitoletic acid* and *oleic acid.* When there ae 2,3,4 or more double bonds in the carbon chain with the consequent absence of 2,6,8 or more hydrogen atoms, such fatty acids are *polyunsaturated fatty acids*. The general formula is CnH_{2n}—2 COOH or CnH_{2n}—3 COOH. Examples are *linoleic acid* and *archidonic acids*.

Linoleic, linolenic and archidonic acid are often termed as *essential fatty acids*.

3. Waxes

Wexes are another class of simple lipis containing one molecule of fatty acid and one high molecular weight alcohol. The constituent acids and alcohols have usually 24-36 carbon atoms. The waxes has high melting points. They are chemically inert as they do not have double bonds in their hydrocarbon chains and are highly insoluble in water. They serve as protective converings on leaf surface (plants). *Bees wax* is an ester of palmitic acid with myricyl alcohol ($C_{30}H_{61}OH$) and *spermaceti* from the sperm whale is an ester of palmitic acid with cetyl alcohol ($C_{16}H_{33}OH$).

II. Compound Lipids

These are esters of fatty acids and alcohol with additional compounds, such as phosphoric acid, sugars, proteins etc. These are classified as follows:

(i) Phospholipids

Lipids containing phosphorous are phospholipids. They also have nitrogen containing bases and other substituents. The phospholipids are abundant in brain and nervous tissues. The different types of phospholipids are:

(a) *Phosphatidic acids*. These are compounds consisting of glycerol, two fatty acids and a phosphate group.

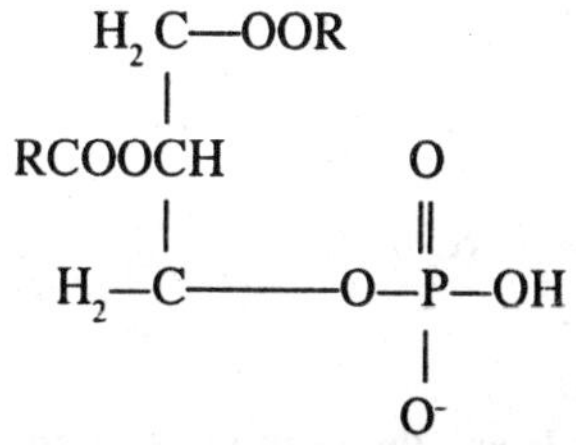

```
  H2C—OOR
    |
RCOOCH          O
    |           ||
  H2—C————O—P—OH
                |
                O-
```

Here R_1 and R_2 represent the residues of the molecules of fatty acids. The phosphatidic acids do not found in any great quantity in tissues but these are important as in intermediate biosynthesis of triglycerides and other phospholipids.

(b) *Lecithins*. Lecithins are choline esters of phosphatidic acid. On hydrolysis they yield one molecule of glycerol, 2 molecules of fatty acids and one molecule of phosphoric acid to which a nitrogenous base choline is attached.

$$
\begin{array}{l}
CH_2.O.CO.R_1 \\
| \\
CHO.CO.R_2 \\
| \qquad\qquad\quad O^- \\
\qquad\qquad\quad\;\; | \\
CH_2O \text{ —— } OP\text{—}O\text{—}(CH_2)_2\text{—}N^+ \; (CH_3)_3 \\
\qquad\qquad\quad\;\; \| \\
\qquad\qquad\quad\;\; O
\end{array}
$$

The fatty acids are palmitic, stearic, oleic, linolic and arachidonic acids. Lecithins are yellowish grey solids, soluble in ether and alcohols but insoluble in acetone. Lecithins get broken down by the enzyme lecithins to lysolecithin. This product (lysolecithin) has the ability to hemolyze the red blood corpuscles. Enzyme lecithinase occurs in the venom of snake cobra. and bee.

Lecithin is an important constituent of lipoproteins. Egg-yolk is a rich source of lecithin which also has an important role in fat metabolism in the liver.

(c) *Cephalins*. They resemble lecithins in most properties. The fundamental difference between the lecithins and cephalins is the nature of nitrogenous base. Cephalins contain ethanolamine, colamine and sometimes serine in place of choline.

$$
\begin{array}{l}
CH_2O.CO.\; R_1 \\
| \\
CHO.CO.R_2 \\
| \\
| \qquad\quad\;\; O \\
| \qquad\quad\;\; | \\
CH_2\text{—}O\text{—}P \text{ —}O\text{—}(CH_2)_2 \; N + H_3 \\
\qquad\qquad\;\; \| \\
\qquad\qquad\;\; O
\end{array}
$$

(d) *Plasmalogens*. These phospholipids are abundant in brian and muscle. They resemble lecithins and cephalins in structure but possess an aldehyde group in place of one of the fatty acids in typical phospholipid molecule.

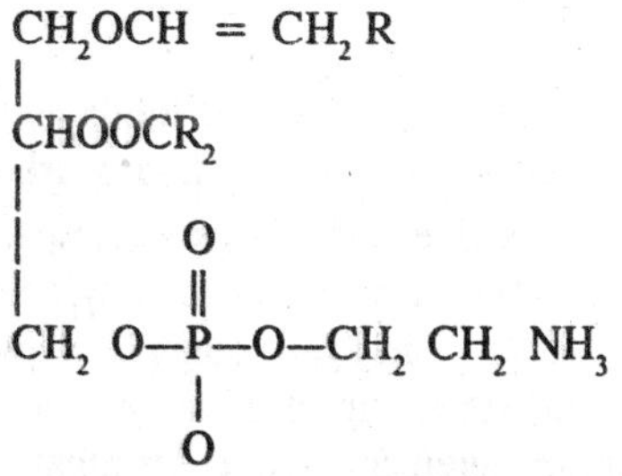

$$
\begin{array}{l}
CH_2OCH = CH_2\,R \\
| \\
CHOOCR_2 \\
| \qquad\quad O \\
| \qquad\quad \| \\
CH_2\ O{-}P{-}O{-}CH_2\ CH_2\ NH_3 \\
\qquad\qquad | \\
\qquad\qquad O
\end{array}
$$

(e) *Phosphoinositides.* Phosphoinositides contain hexahydric alcohol inositol. They can be either monophosphoinositides or diphosphoinositides. An another name lipoinositol was also proposed for these substances. They occur in animal and plants but diphosphoinositides have been reported from brain tissue only.

(f) *Phosphosphingosides.* Phosphosphingosides have been found to occur in high concentration in brain, nerves, lungs and spleen tissue. These contain a nitrogenous base sphingosine or its derivative which remains attached to long-chain fatty acid by its amino group. On hydrolysis they yield fatty acids, phosphoric acid, choline and sphingosine.

(ii) Glycolipids

The glycolipids occurs in brain tissue and in the myelin sheath of nerves, lungs, kidneys, spleen, liver, retina, egg-yolk etc. The glycolipids include two major catagories, cerebrosides and gangliosides.

(a) *Cerebrosides.* They occur in the brain and myelin sheath of nerves. they are based on sphinge mine and have in addition a fatty acid and a monosaccharide sugar but no phosphoric acid or glycerol. Individual cerebrosides are differentiated by the types of fatty acids in the molecule. the important types are:

Kerasin containing the saturated lignoceric acid, $CH_3\ (CH_2)_{22}\ COOH$.

Cerebron containing the cerebronic acid, $CH_3\ (CH_2)_{21}\ CHOH\ COOH$.

Nervon containing the nervonic acid $CH_3\ (CH_2)_7\ CH = CH\ (CH_2)_{13}\ COOH$.

Oxynervone containing hydroxy derivatives of nervonic acid $CH_3\ (CH_2)_7\ CH{=}CH\ (CH_2)_{12}\ CHOH\ COOH$.

(b) *Gangliosides.* These contain N-acetylneuraminic acid (sialic acid), fatty acids, sphingosine and 3 molecules of hexose.

III. Derived Lipids

These lipids include hydrolytic products of lipids as well as other lipid like compounds like sterols, carotenoids, hydrocarbous etc.

The *sterols* are solid wax-like substances chemically, they are alcohols and occur either as such or as esters of fatty acids. They are highly soluble in fat solvents. They all contain a cyclopentano-phenanthrene (sterane) nucleus (made up of three cyclohexane rings, in the phenanthrane type of arrangement and a terminal cyclopentane ring).

Cholesterol is animal origin occuring in bile, brain, spinal cord etc. It is obtained form human gall bladder stones which are deposited in the bile duct.

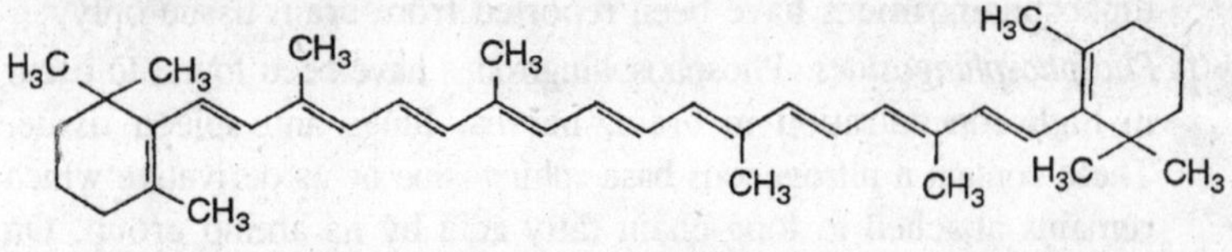

Ergosterol. It is similar to 7-dehydrocholesterol but differs in the side chain. It can be converted into vitamin D on exposure to the ultra violet light and, therefore, also called provitamin D.

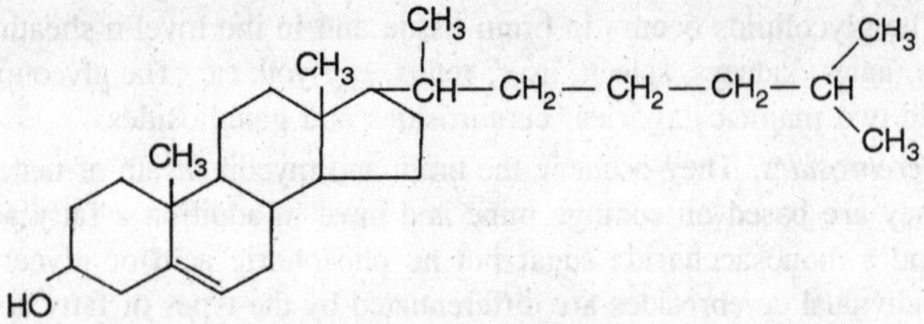

Carotenoids. These are also included in lipids because of their solubility in fat solvent. Carontenes consists of carbon and hydrogen only where as xanthophyll contain oxygen in addition. Vitamin A is derived from carotene. *Estrogens, progesterone, testosterone* and *anderosterone* are steroidal hormoens.

Prostaglandins

Prostaglandins are derivatives of fatty acids. They were first discovered in human seminal fluid secreted by prostate gland, hence the name. It has been shown now that prostaglandins are synthesized and released by many other tissues such as kidneys, testis, placenta, lungs, liver, uterus, gastrointestinal tract, brain and heart.

Table 2.4. Showing Elements that Occur in Protoplasm

S.No.	*Name of elements*	*Approximate percentage*	*Function*
Major Elements			
1.	Carbon	18	Forms backbone of all organic molecules.
2.	Hydrogen	10	Present in most organic com-pounds and major components of water.
3.	Oxygen	65	Cellular respiration; in organic compound and component of water.
4.	Nitrogen	03	Major components of all the amino acids, proteins and nucleic acids.
Tracer Elements			
5.	Sodium	0.2	For water balance, conduction of nerve impulse.
6.	Potssium	0.5	Conduction of nerve impulse, muscular contraction.
7.	Phosphorus	1.0	Nucleic acids formation; bones formation and in energy transfer.
8.	Magnesium	0.1	Constituent of certain enzymes (ATPase).
9.	Calcium	1.5	Blood clotting, muscle contraction, bones and teeth formation.
10.	Sulfur	0.3	Constituent of most proteins.
11.	Chloride	0.1	Negative ion of interstitial fluid.
12.	Iron	0.01	Component of haemoglobin and certain enzymes.

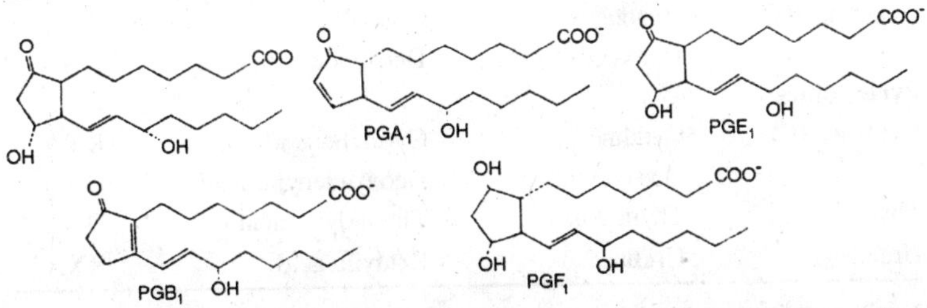

The prostaglandins are 20 carbon fatty acids including a five membered ring in their molecular structure. Different prostaglandins differ with one another in the number and position of double bonds and hydroxyl group substituents. Prostaglandins PGE, PGE_1, PGE_3, PGF_{1a}, PGF_{2a} and PGF_{3a} are considered as *primary prostaglandins*. Most of their action appear to be binding of hormones to membrane.

Recently, new type of prostaglandin has been isolated from human seminal fluid which has been designated as PGX[2].

Amino Acids

There is a common plan of construction for the thousands kinds of proteins in living systems. The 20 kinds of naturally occurring amino acid monomers are strung together in unbranched, linear polymer chains of proteins. These are the 20 amino acids specified in the genetic code that is universal to all organisms. Some other kins of amino acids are also found in cells, but they are either degradation products or residues that have been modified form oen of the 20 commonly occurring amino acids after this latter has been inserted into the polymer chain. Hydroxyproline is a major amino acid constituent of collagen in connective tissue, but proline residues are initially included in the protein and are converted to hydroxyproline after polymerization. Hydroxyproline is not one of the encoded acid, but proline is. Many proteins contain fewer than 20 kinds of amino acids. The relative proportions and the absolute number of the amino acid repertory vary from one protein to another, as a reflection of the specific in formation in genes, which are the blue prints for protein construction.

Table 2.5. Showing nomenclature of nucleic acids and their constituent units.

Base	*Nucleoside*	*Nucleotide*	*Nucleic Acid*
Purines			
Adenine (A)	Adenosine	Adenylic acid	RNA
	Deoxyadenosine	Deoxyadenylic acid	DNA
Guamine (G)	Guanosine	Guanylic acid	RNA
	Deoxyguanosine	Deoxyadenylic acid	DNA
Pyrimidines			
Cytosine (C)	Cytidine	Cytidylic acid	RNA
	Deoxycytidine	Deoxyadenylic acid	DNA
Thymine (T)	Thymidine	Thymidylic acid	DNA
Uracil (V)	Uridine	Uridylic acid	RNA

Nucleotides and Nucleic Acids

Nucleotides are involved in at least two major cellular functions: (1) they are monomeric units from which DNA and RNA polymers

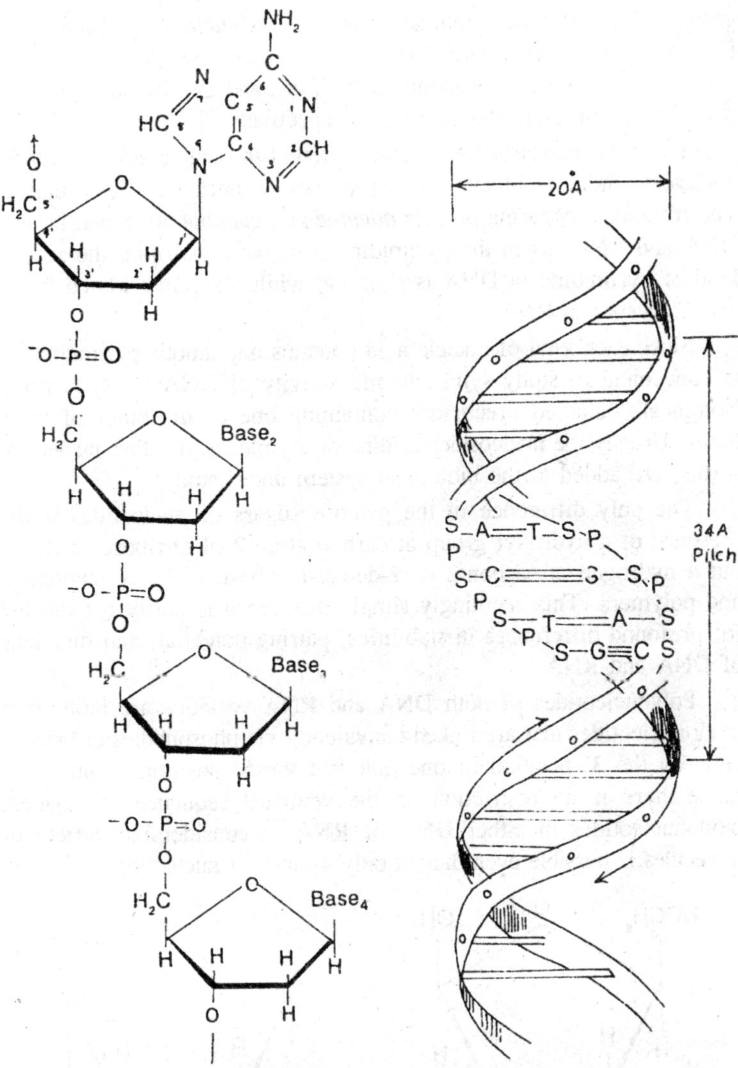

Fig. 2.6. The molecular organization of a backbone of DNA showing phosphodiester and glycosidic bonds.

Fig. 2.7. DNA and its paired bases.

are constructed, and (2) they act as agents in certain energy-transferring reactions during metabolism. A mononucleotide is made up of one nitrogen-containing organic base, one pentose sugar, and one phosphate residue derivative from phosphoric acid. When there is no phosphate group, the sugar-base combination is called a *nucleoside* (Table 2.5). For this reason, *nucleotides* (phosphate-sugar-base) are also called nucleoside phosphates. Nucleoside mono-, di-, and tri- phosphates contain one, two or three phosphate groups respectively.

The nitrogenous bases commonly found in nucleic acids and their nucleotide building blocks are derivatives of purine and pyrimidine. The commonly occuring purines *adenine* and *guanine* are found in both DNA and RNA, as in the pyrimidine compound *cytosine,* the second kind of pyrimidine in DNA is *thymine,* while its demethylated form, *uracil,* occurs in RNA.

Since each kind of nucleic acid contains one unique pyrimidine, it is convenient to study synthesis and activity of DNA or RNA using isotopically-labelled precursors containing one or the other of these bases. Usually the nucleosides uridine or thymidine, or other nucleotide forms, are added to the biological system under study.

The only difference in the pentose sugars of nucleotides is the presence of a hydroxyl group at carbon atom 2 of D-ribose in RNA, but a hydrogen at carbon-2 of 2-deoxy-D-ribose in DNA monomers and polymers. This seemingly simple difference is partly responsible for profound differences in stabilities, pairing potential, and functions of DNA and RNA.

Polynucleotides of both DNA and RNA varieties are built from mononucleotides that are linked convalently via phosphodiester bridges between the 3' position of one unit and the 5' position of hte next. Since there is no restriction on the ventrical sequence of adjacent mononucleotides in either DNA or RNA, a considerable variety of molecules is possible even though only 4 kinds of nucleotide monomers

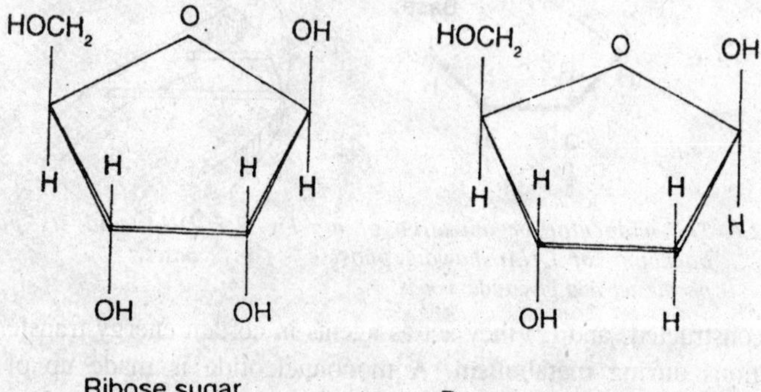

Fig. 2.8. Pentose sugar.

(one for each of the four kinds of bases in the combination with sugar and phosphate) are used in polymer construction. The theoretical variety is calcualted as 4”, where 4 is the number of different kinds of nucleotides, and *n* is the number of monomers in the polymer. For a molecule made of only 75 monomeric units, as in some of the smallest RNAs, there may be 4^{75} different arrangements of the constituent units. Each arrangement theoretically constitutes a moleule of different specificity. Where the average gene may include about 500 nucleotides in a DNA sequence, 4^{500} different sequences are theoretically possible and, therefore, that many different and specific genes. Despite the apparently meager number of monomer types an astronomically high number of possible genes can be constructed.

Such variety can easily account for all past and present life forms.

The DNA Double Helix

DNA molecules usually have regular helical configurations because most DNA molecules consists of two *complementary polynucleotide strands*. The two strands are held together by *hydrogen bonds* between complementary pairs of purines and pyrimidines. Adenine always bind with thymine while guanine always binds with cytosine. This repeated

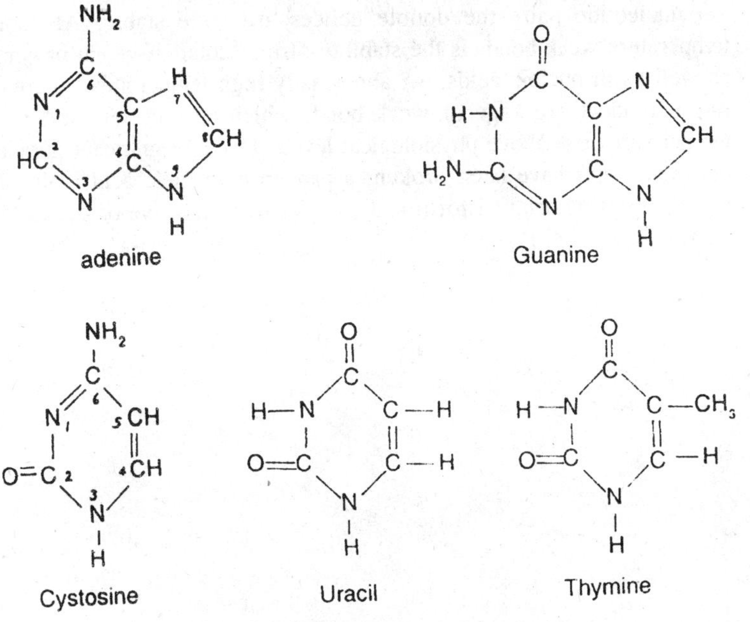

adenine

Guanine

Cystosine

Uracil

Thymine

Fig. 2.9. Various heterogenous bases.

hydrogen bonding within the double helix structure and bonding between virtually all the surface atoms in the sugar and phosphate groups with water molecules serve to stabilize the structure.

Since the purine-pyrimidine pairs are found in the center of the molecule, their flat surface can stack on top of each other and thereby limit their contact with water. In double-helical molecules, a regular structure is possible because the comple-mentary base pairs are exactly the same size. Single polynucleotide chains could not have a regular backbone structure because pyrimidines are smaller than purines, which would cause the angle of helical rotation to vary with the sequence of bases.

DNA double helix molecules are very stable at physiological temperatures because: (1) disruption of the double helix breaks hydrogen bounds and brings hydrophobic purines and pyrimidines into contact with water, which is energetically unsatisfactory; and (2) there are many weak bonds within the DNA molecule, arranged so that most of them cannot break without many others breaking at the same time. Even though some hydrogen bonds may be broken by thermal motion, hydrogen bonds in the rest of the molecule remain intact and the molecule does not fall apart. In fact, when held together by more than ten nucleotide pairs the double helices are quite stable. At room temperature weak bonds is the stability of molecular shape, in proteins as well as in nucleic acids. At abnormally high temperatures there is more frequent breakage of weak bond, which become less stable as temperatures rise above physiological levels. Once a significant number of weak bonds have been broken, a protein or nucleic acid molecule usually loses its original form and changes to an inactive or denatured form.

3

MOLECULES IN LIFE

One of the oldest arguments in the history of biology concerned the basic nature of living things. Many biologists in the past thought that the substances composing living things are fundamentally different from those in nonliving things. They thought there is a special life chemistry, governed by its own set of laws different from those governing nonliving matter. They also proposed that there is a special force or energy, "the vital force," that is unique to life. Despite all efforts to detect it, however, no such vital force has ever been found, and chemical laws have been found to operate the same whether the chemical processes occur in living or nonliving things.

Today, biologists recognize that while all matter—living and nonliving—consists of atoms and molecules, many of the molecules in living organisms are larger and more complex than those commonly found in nonliving systems. As we will see, these relatively large and complex molecules are arranged into highly organized systems such as membranes and cells. Because of the high degree of organization among their molecules, living things have properties not found in nonliving systems.

In order to be able to explore the properties of living things, we need to understand some characteristics of the substances that constitute them.

ELEMENTS AND COMPOUNDS

All matter is composed of basic substances called *elements*. An element cannot be broken down into simpler units by chemical reactions; it contains only one kind of atom. An *atom* is the smallest characteristic unit of an element.

Atoms of most elements do not occur in isolation. Usually they are combined with other atoms to form *molecules*. A molecule is a combination of like or different atoms. For example, we do not find individual oxygen atoms in the atmosphere. Rather, oxygen in air is in the form of oxygen molecules, each of which contains two oxygen atoms.

There are ninety-two naturally occurring elements and a number of artificial elements that can be made in laboratories. Each element has a symbol consisting of a one- or two-letter abbreviation of its English or Latin name. For example, the symbols for hydrogen (H) and oxygen (O) are derived from their English names, while the symbols for sodium (Na) and potassium (K) are derived from their Latin names, *natrium* and *kalium*.

A *compound* is a substance that can be split into two or more elements. Water is a compound because it can be split into its components, hydrogen and oxygen. The *formula* of a compound gives information about the kinds and numbers of atoms that make up each molecule of that compound. A formula contains the symbols (abbreviations) of the kinds of atoms in each molecule and subscripts that indicate the number of each kind of atom in the molecule. For example, the formula for water, H_2O, indicates that a water molecule contains two hydrogen atoms and one oxygen atom; and a molecule of the sugar glucose, $C_6H_{12}O_6$, contains six carbon atoms, twelve hydrogen atoms, and six oxygen atoms.

Atomic Structure

While atoms are the smallest units that possess all the properties of the particular elements, each atom is composed of still smaller particles.

Protons, Neutrons, and Electrons

There are several kinds of subatomic particles, but only three are crucial to the properties of elements we will consider here: protons, neutrons, and electrons. *Protons* and *neutrons* are located in a central area of the atom called the *nucleus*. *Electrons* are located at various distances from the nucleus. Protons have a positive electric charge, and electrons have a negative charge; thus, protons attract electrons. Neutrons have no electric charge.

While protons and neutrons have nearly the same mass, their mass is much greater than the mass of electrons (more than 1,800 times as great). Chemists customarily use the *Atomic Mass Unit* (AMU), or *dalton*, to describe the mass of subatomic particles. Both the proton and the neutron have masses of approximately one dalton.

Each atom of an element has a characteristic number of protons in its nucleus, and this number is known as the element's *atomic number*. For example, because a carbon atom has six protons in its nucleus, its atomic number is six. You can see the atomic numbers of some biologically important elements in table 3.1.

Table 3.1. Some elements important in biology.

Element	*Symbol*	*Atomic number*	*Atomic weight*
Hydrogen	H	1	1.01
Carbon	C	6	12.01
Nitrogen	N	7	14.01
Oxygen	O	8	16.00
Phosphorus	P	15	30.97
Sulfur	S	16	32.06
Sodium	Na	11	23.00
Magnesium	Mg	12	24.31
Chlorine	Cl	17	35.45
Potassium	K	19	39.10
Calcium	Ca	20	40.08
Iron	Fe	26	55.85
Copper	Cu	29	63.54
Zinc	Zn	30	65.37

Normally, the number of electrons outside the nucleus of the atom is equal to the number of protons in the nucleus. Because the positive charges of the protons equal the negative charges of the electrons, the charges cancel one another and the atom is neutral.

The *mass number* of an element indicates how many protons and neutrons are in the nucleus of one of its atoms. For example, most carbon atoms contain six protons and six neutrons. Carbon's mass number, 12, is symbolized with a superscript numeral preceding the element's symbol: ^{12}C (read "carbon-12"). The *atomic weight* is the *actual measured weight* of an element and is usually very close to, but not exactly the same as, the mass number.

Isotopes

Use of the word isotope has become common in our society. We hear almost daily about the use of isotopes, particularly radioactive isotopes (or simply "radioisotopes") in medical diagnosis and cancer treatment. We also hear these terms in discussions of nuclear power plants and nuclear waste disposal. What is meant by the term "isotope," and what are radioactive isotopes?

All atoms of a given element have the same number of protons, but different atoms of that element can have different numbers of neutrons. Recall that the mass number of an element is equal to the number of protons plus the number of neutrons in each of its atomic nuclei. The atoms of different forms, or *isotopes*, of a given element have different numbers of neutrons and different mass numbers.

As we saw above, the most common form of carbon atom, with six protons and six neutrons in its nucleus, has a mass number of 12 (^{12}C). A carbon isotope with six protons and seven neutrons is ^{13}C, while a carbon isotope with six protons and eight neutrons is ^{14}C.

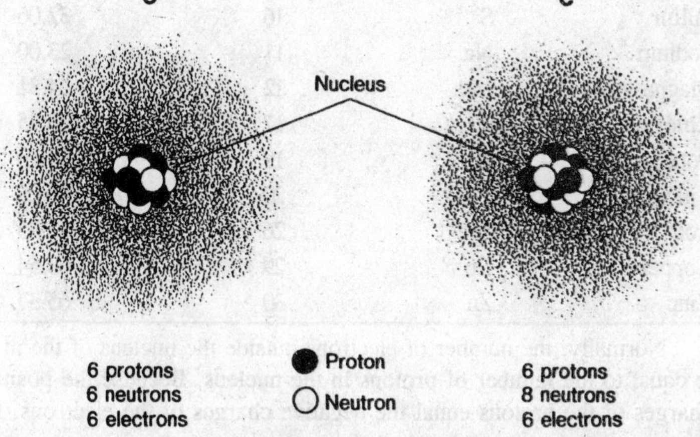

Fig. 3.1. Nuclei of carbon isotopes differ only in the number of neutrons present.

Some isotopes are stable, while others are unstable and tend to decay (break down) by spontaneously emitting small particles. These unstable isotopes are ***radioactive***. ^{12}C and ^{13}C are stable isotopes, while ^{14}C is radioactive.

Radioactive isotopes of some elements emit particles with a great deal of energy. Some of the radioisotopes produced in nuclear reactors emit very high energy radiation and are extremely dangerous to handle and difficult to store safely.

Certain isotopes that emit lower-energy radiation are very useful in biological research and medical diagnosis. Radioactive isotopes can be used as chemical "labels" to mark certain compounds in cells or tissues. Because the different isotopes of an element react chemically in the same way, radioactive isotopes may be substituted for stable isotopes to make chemical substances under observation easier to detect.

Electrons and Orbitals

Although electrons have a much smaller mass than protons and neutrons, they are by no means less important in determining the chemical properties of atoms. The location of electrons in relation to the nucleus has a great influence on the way atoms and molecules react with each other.

Because electrons are continually in motion around the nucleus in every atom, we cannot say exactly where a particular electron is located at any one moment. We can say, however, where electrons are most likely to be located most of the time. We call the space in which an electron is located 90 percent of the time the electron's *orbital*.

Each orbital can contain a maximum of two electrons, and orbitals are grouped into *shells*. The electrons located in the various shells have different energy levels. The shell closest to the nucleus has the lowest energy level, and each shell beyond has a progressively higher energy level as the distance between nucleus and shell increases.

The first shell, the one closest to the nucleus, contains only one orbital (a spherical one), and thus can contain only two electrons. The second shell contains four orbitals, one spherical and the other three shaped like dumbbells. This second shell, then, can contain up to eight

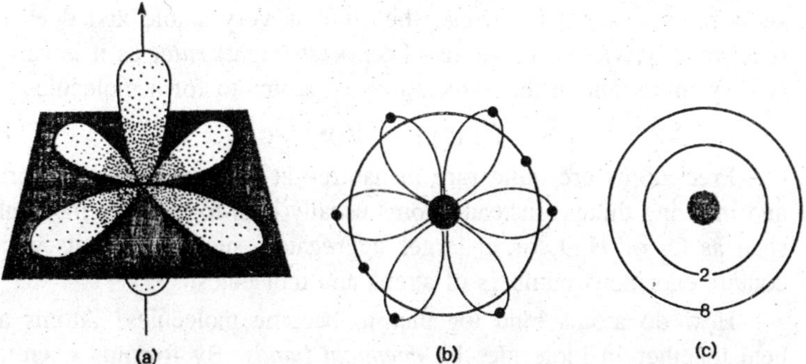

Fig. 3.2. Electron orbitals. (a) The three dumbbell-shaped orbitals of the second shell are at right angles to each other. (b) All four orbitals of the second shell on a flat plane. (c) A simpler way of representing the orbital shells with the number of electrons contained in each

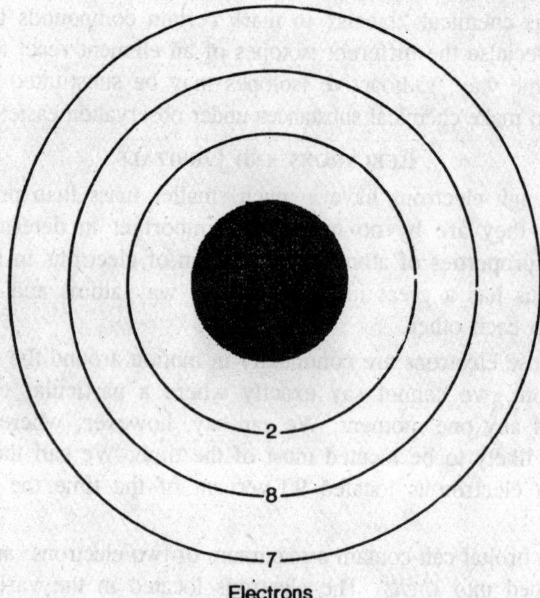

Fig. 3.3. Electron orbital shells of a chlorine atom.

electrons. Although the third shell and those beyond can hold more than eight electrons, these shells are in a particularly stable condition when they do hold exactly eight electrons.

Under normal circumstances, electrons will be in the innermost shells in which orbital space is available. In this way, each lower-energy shell is filled before the next shell is. When an atom's outermost shell contains eight electrons, the atom is very stable and does not react readily with other atoms. This *octet* (eight) *rule*, as it is called, is very important in the associations of atoms to form molecules.

Chemical Bonding

Free atoms are quite rare in nature—at least in the earth's crust and in living things. Instead, atoms usually bind together in molecules such as O_2 or H_2O, or in larger aggregates, such as crystals which contain enormous numbers of atoms and molecules.

How do atoms bind together to become molecules? Atoms are held together in molecules by *chemical bonds*. By forming chemical bonds, atoms tend to achieve a more stable electron configuration than they have by themselves. A stable molecule is formed when the uniting atoms can reshuffle their electrons so that each atom has a complete, or at least more stable, outer shell.

Atoms combining into molecules rearrange their electrons through different means. In *ionic bonding*, electrons are *transferred* from one atom to another. In *covalent bonding*, pairs of electrons are *shared* between atoms.

Ionic Bonding

To illustrate ionic bonding, let us examine the reaction between sodium (Na) and chlorine (Cl) in which sodium chloride (NaCl), common table salt, is formed. Sodium's atomic number is eleven—a sodium atom has eleven protons in its nucleus and eleven electrons orbiting around the nucleus. The inner shell, containing two electrons, is complete, as is the second shell with eight electrons. But the third shell of a sodium atom contains only one electron. An atom with only one electron in its outer shell tends to be an *electron donor*—that is, it tends to get rid of its lone electron.

Chlorine has an atomic number of seventeen, which means that a chlorine atom has two electrons in its first shell, eight in its second shell, and seven in its outer shell. Because it needs only one electron to complete its outer shell (remember the octet rule), a chlorine atom tends to be an *electron acceptor*. When a sodium atom and a chlorine atom come together, an electron is transferred from the sodium atom

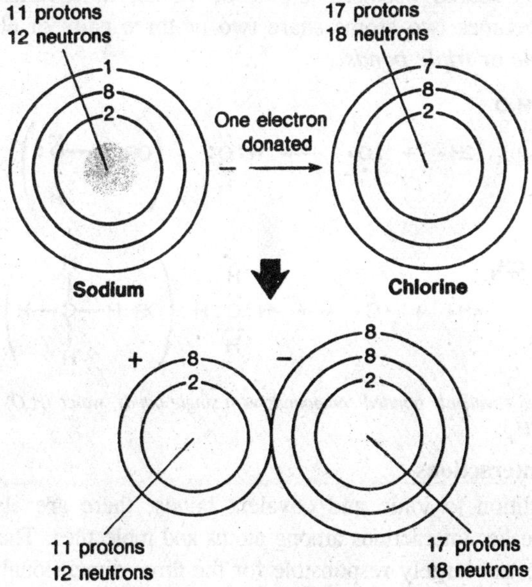

Fig. 3.4. Ionic bonding

to the chlorine atom. But this electron transfer changes the balance between the protons and electrons within each of the two atoms. The sodium atom ends up with one more proton than it has electrons, and the chlorine atom with one more electron than it has protons. The sodium atom is left with a net charge of +1 (symbolized Na^+), while the chlorine atom's net charge is -1 (symbolized Cl^-). Charged atoms are called *ions* and are attracted to one another by their opposite charges. Thus, an ionic compound such as sodium chloride (NaCl) is held together by the attraction between its positive and negative ions; that is, by an *ionic bond*.

The ionic bonds of sodium chloride are quite strong when the salt exists as a dry solid, but when NaCl is dissolved in water, it separates into Na^+ and Cl^- ions. Ionic compounds are most commonly found in this *dissociated* (*ionized*) form in the watery environment inside living things.

Covalent Bonding

In *covalent bonds* electrons are not donated from one atom to another, but are shared between atoms. Thus, a covalent bond consists of a pair of electrons that are shared between two atoms and that occupy two stable orbitals, one of each atom. If a single pair of electrons is shared between a pair of atoms, it is called a *single bond*. Sometimes two atoms share two or three pairs of electrons to form *double* or *triple bonds*.

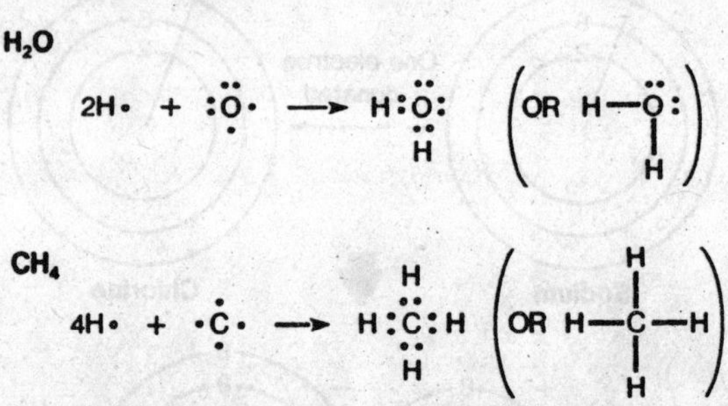
H2O
2H• + :Ö• —► H:Ö:
H
OR H—Ö:
H
CH4
4H• + •C• —► H:C:H
H
H
OR H—C—H
H
H

Fig. 3.5. Two covalently bonded compounds with single bonds, water (H_2O) and methane (CH_4).

Weaker Interactions

In addition to ionic and covalent bonds, there are also several types of weaker interactions among atoms and molecules. These weaker interactions are largely responsible for the three-dimensional shapes of many large biological molecules, such as proteins, and for some of

CO_2 $\quad$ O=C=O

(a)

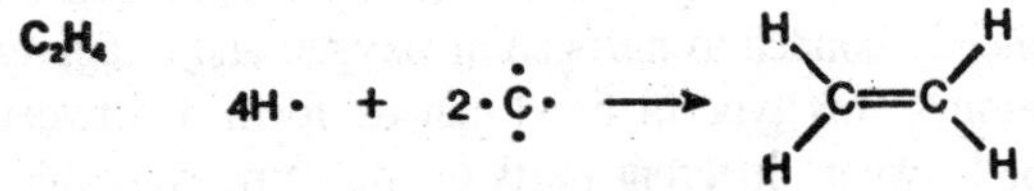

C_2H_4 $\quad$ $4H\cdot + 2\cdot\dot{C}\cdot \longrightarrow$ H₂C=CH₂

(b)

C_2H_2 $\quad$ $2H\cdot + 2\cdot\dot{C}\cdot \longrightarrow$ H—C≡C—H

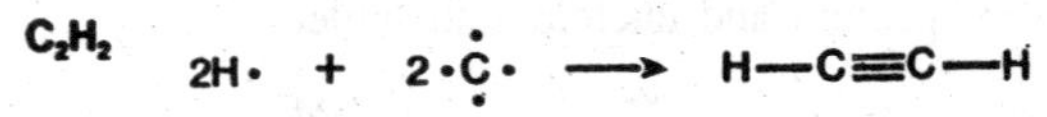

(c)

Fig. 3.6. Some compounds with multiple covalent bonds.

their functional properties. In some molecules formed by covalent bonds, the electrons tend to be attracted more toward one atom than the other. This unequal amount of attraction among the electrons is called *polarity*. Polarity results in molecules that have a small positive charge at one end and a small negative charge at the other. Because of these slight charges at their ends or poles, polar molecules are attracted to

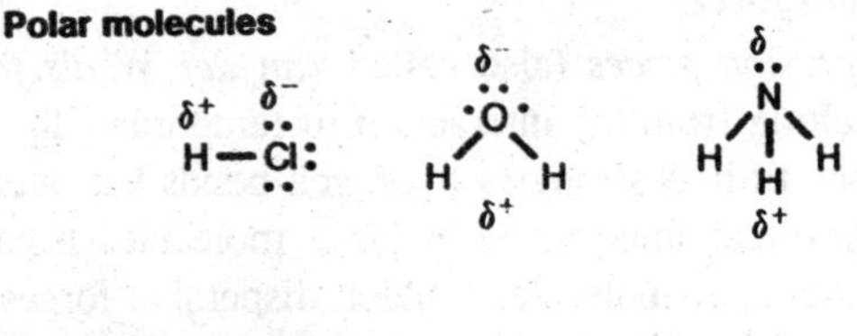

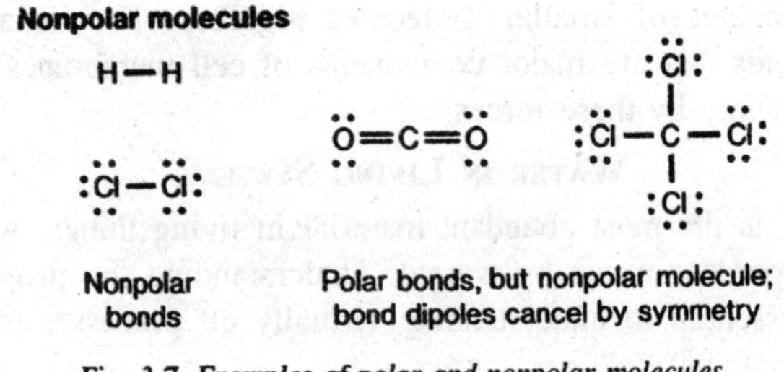

Nonpolar molecules

H—H

:Cl—Cl:

O=C=O

:Cl: / :Cl—C—Cl: / :Cl:

Nonpolar bonds

Polar bonds, but nonpolar molecule; bond dipoles cancel by symmetry

Fig. 3.7. Examples of polar and nonpolar molecules.

each other. The weak electrical attractions between opposite charges on polar molecules are known as *dipole forces*.

Hydrogen bonding

An example of this attraction between polar molecules is *hydrogen bonding*. Hydrogen bonding occurs between a hydrogen atom that is already covalently bonded to nitrogen or oxygen and a *different* nitrogen or oxygen atom. This type of bond can be formed between different molecules or between different parts of the same molecule.

Both types of hydrogen bonding are important in biological processes. For instance, hydrogen bonding between different molecules is crucial to the role of water in living systems. Hydrogen bonding within molecules is especially important in the threedimensional organization of protein and nucleic acid molecules.

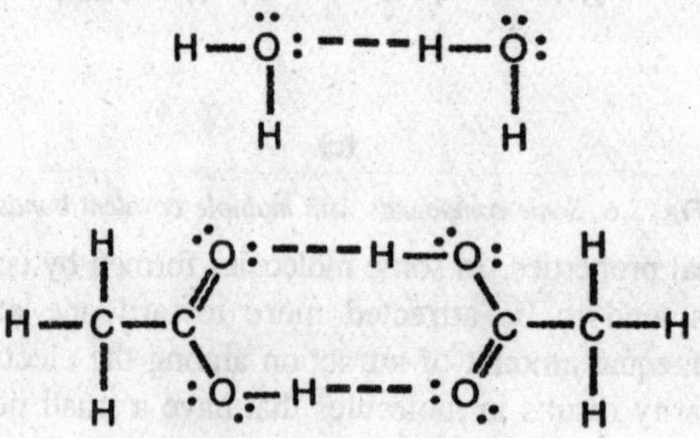

Fig. 3.8. Hydrogen bonding.

London dispersion forces

London dispersion forces (also called *van der Waals forces*) are weak forces resulting from the interaction of temporary dipoles. They are only about one-fifth as strong as hydrogen bonds but, weak as they are, they assume great importance in large molecules because there are so many of them per molecule. London dispersion forces also help hold aggregates of smaller molecules together. For instance, the phospholipids that are major components of cell membranes are held together mainly by these forces.

WATER IN LIVING SYSTEMS

Water is the most abundant material in living things, which are 60 to 95 percent water by weight. Understanding the properties of water is essential to understanding virtually all processes that occur in living systems.

Some Properties of Water

The structure of the water molecule is nonlinear (bent) with a bond angle of 104.5° between the HOH (hydrogen-oxygen hydrogen) atoms. Water is a polar molecule, with the oxygen atom forming the

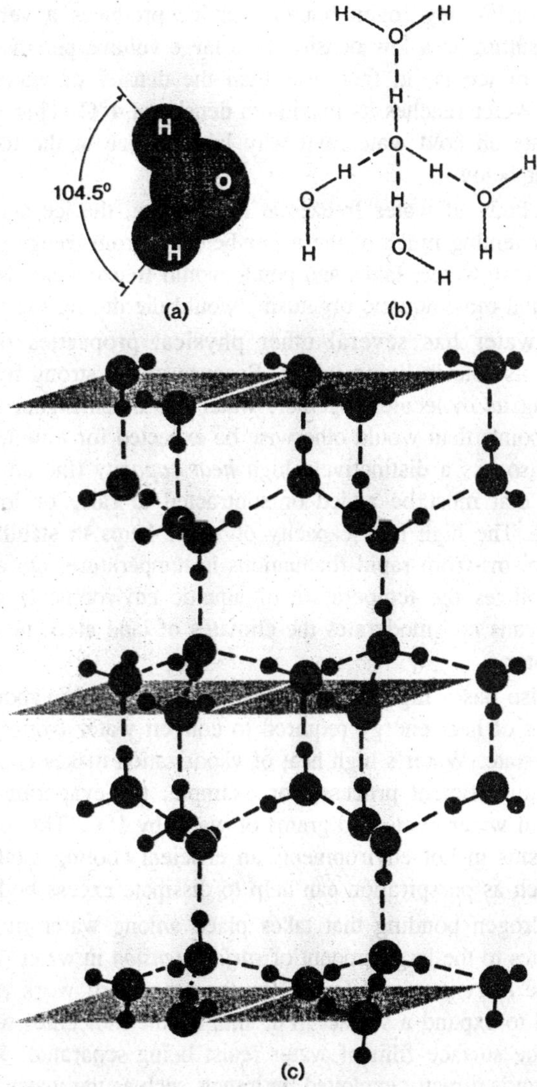

Fig. 3.9. (a) Structure of a water molecule. (b) Tetrahedral hydrogen bonding of water molecules in ice. (c) The open structure of ice.

negative end and the hydrogen atoms the positive end. Therefore, hydrogen bonding occurs between water molecules. We can see the importance of hydrogen/ bonding in water most strikingly by examining the structure of ice. In ice, each oxygen atom is surrounded tetrahedrally by four hydrogen atoms, two close (covalent bonds) and two distant (hydrogen bonds). Hydrogen bonding in ice produces a very open structure, resulting in a low density (or a large volume per molecule). The density of ice is, in fact, less than the density of water in its liquid state. Water reaches its maximum density at 4°C. This explains why ice floats on cold water and why lakes freeze at the top first, leaving water below.

When a body of water freezes at the surface, the ice acts as an insulator, preventing much of the water below it from freezing. If ice were denser than water, lakes and ponds would freeze solid, from the bottom up, and most aquatic organisms would die during the winter.

Liquid water has several other physical properties that are important to its role in living things. Because of the strong hydrogen bonds holding its molecules together, water has much higher melting and boiling points than would otherwise be expected for a molecule of its size. It also has a distinctively high *heat capacity* (the amount of heat energy that must be added or subtracted to raise or lower its temperature). The high heat capacity of water helps to stabilize and protect organisms from rapid fluctuations in temperature. On a larger scale, it stabilizes the temperature of aquatic environments such as lakes and oceans and moderates the climates of land areas near large bodies of water.

Water also has a high *heat of vaporization.* Heat of vaporization is the amount of heat energy required to convert water from a liquid to a gaseous state. Water's high heat of vaporization makes cooling by evaporation an efficient process; for example, the evaporation of a single gram of water cools 540 grams of water by 1°C. This property gives organisms in hot environments an efficient cooling method, as processes such as perspiration can help to dissipate excess body heat.

The hydrogen bonding that takes place among water molecules also contributes to the large amount of *surface tension* in water. Surface tension is the *force per unit length* (or the amount of work per unit area) needed to expand a surface. Put simply, the molecules of water making up the surface film of water resist being separated. Water's surface tension is directly exploited by insects such as the water strider, which can "skate" on a water surface, its small weight supported by surface tension. The mutual attraction of water molecules is also important in many other biological processes, including the movement of fluids in plants.

Water as a Solvent

Major parts of all living things consist of watery solutions. In a *solution*, there is a uniform mixture of the molecules of two or more substances. In living things, water is by far the most important *solvent*, the dissolving substance that is present in the greatest amount. Thus, molecules of various *solutes* (dissolved substances) are dispersed among molecules of water in solutions in living things.

How does water function as a solvent, and what determines which substances will or will not dissolve in water? Water's properties as a solvent relate to the polarity of water molecules. Recall that one end (pole) of each water molecule has a slight positive charge, and the other a slight negative charge. Many molecules that are important in living things are polar or have polar regions. The positive regions of such molecules are attracted to the negative ends of water molecules and vice versa, and water molecules are similarly attracted to them. The attractive forces involved are called *hydrophilic* ("water-loving") interactions. Because of these hydrophilic interactions with water molecules, such polar molecules become interspersed among water molecules. Sugars, salts, some alcohols, and ammonia are examples of water-soluble substances.

Other molecules, however, are nonpolar or have large nonpolar parts, and those molecules do not dissolve in water. This is because the forces attracting water molecules to one another are stronger than those acting between water and nonpolar molecules. Thus, the nonpolar molecules are "squeezed out" from between water molecules. Because nonpolar molecules do not interact with water molecules, they cannot become dispersed among water molecules to form a solution. Fats and oils are good examples of such substances. No matter how vigorously you stir oil in water, the oil remains in droplets and does not go into solution. The interactions between water molecules and nonpolar molecules or nonpolar parts of molecules are referrerd to as *hydrophobic* ("water-hating") interactions.

When an ionic compound, such as sodium chloride (NaCl), is dissolved in water, its ions separate because the charged poles of the water molecules attract the ions more than the ions attract each other. Not only are the ions pulled apart in the process, but they become surrounded by layers of water molecules called *hydration spheres*. The

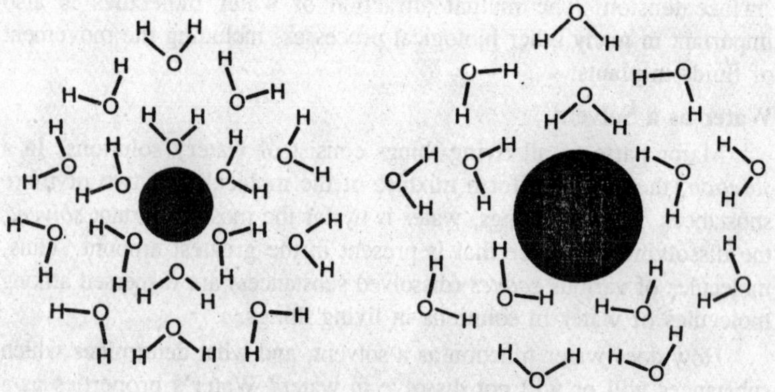

Fig. 3.10. Hydration of ions, using Na^+ and Cl^- ions as examples.

water molecules around the ions are oriented so that the negative (O) end of each water molecule is adjacent to the positive ion (Na^+). Similarly, the positive (H) end of each water molecule is attracted to the negative ion (Cl^-). Hydration spheres effectively increase the size of ions and affect the movement of ions in biological systems, including the passage of ions through cell membranes. The hydration process also affects the structure of water, because the water molecules in each sphere are oriented in one direction—with either their positive or negative ends attracted toward the ion. Because of their fixed orientation, the water molecules cannot, for example, take part so readily in ice formation. This explains why the higher the salt content of water, the more resistant it is to freezing—solutes depress the freezing point of water.

What happens when a molecule with both a small, polar (hydrophilic) portion and a large, nonpolar (hydrophobic) portion is added to water? When molecules with polar "heads" and long, nonpolar "tails" are mixed with water, they form structures known as *micelles*. In a micelle the nonpolar tails turn toward each other and away from the H_2O molecules. However, hydrophilic attraction does occur between the polar heads and adjacent H_2O molecules. Detergent action is a familiar, everyday example of micelle formation. A detergent disperses nonpolar grease by incorporating the grease molecules into the interior of micelles.

The phenomenon of micelle formation reveals much about the structure of biological membranes because biological membranes are largely made up of molecules that have both hydrophilic and hydrophobic ends.

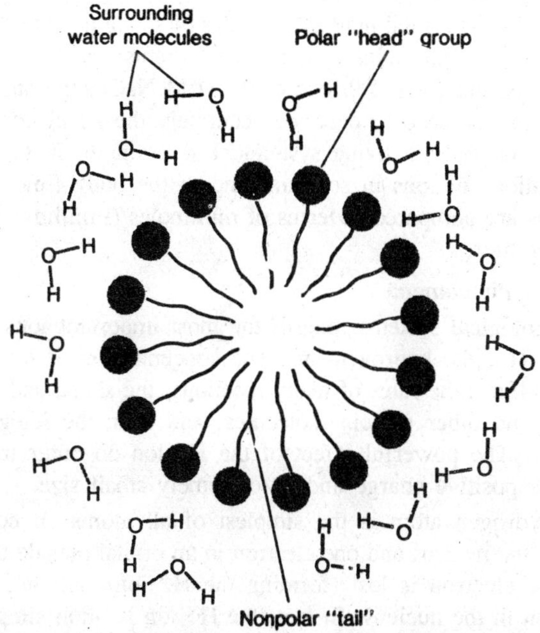

Fig. 3.11. The formation of a micelle by molecules that have a polar "head" and a nonpolar "tail."

Concentrations of Solutions

The properties of a solution depend not only on the natures of the solvent and solute, but also on their concentrations. We can express the concentrations of solutions in several ways. Stating the concentration as a *percentage* (normally, weight of solute to volume of solvent) is one simple approach. Physiological saline solution, for example, is 0.9 percent (weight-to-volume) NaCl in water.

In describing chemical reactions, however, the *mole* is a more useful unit of measurement for the amounts of different substances. A mole is the mass of substance, expressed in grams, that is numerically equal to the *molecular* (or *formula*) *weight*. The formula weight of any molecule is equal to the sum of the atomic weights of the atoms in the molecule. Thus, the formula weight of NaCl is 58.5, and one mole of NaCl is 58.5 grams:

23.0 (atomic weight of Na) + 35.5 (atomic weight of Cl) = 58.5

The mole is useful for expressing quantities of substances. One mole of any substance contains Avogadro's number (6.02×10^{23}) of molecules of the substance.

Molarity (*M*) is the number of moles of solute per liter of solution. Physiological saline solution contains 9 grams of NaCl per liter of solution, and thus, it is a 9/58.5 or 0.154 M NaCl solution. Moles and molarity can be used to describe accurately the small concentrations of many materials in living systems. For example, in table 2.2, the concentrations of ions in seawater and in the body fluids of several organisms are compared in terms of millimoles (1 millimole = 1/1,000 mole) per liter.

Acid-Base Phenomena

In biological systems, one of the most important ions is also the simplest: H^+, the hydrogen ion. The concentration of hydrogen ions strongly affects the rates of many reactions, the shape and function of enzymes and other protein molecules, and even the integrity of the cell itself. The powerful effect of the H^+ ion on other molecules is due to its positive charge and its extremely small size.

A hydrogen atom is the simplest of all atoms. It contains one proton in the nucleus and one electron in an orbital outside the nucleus. When the electron is lost (forming the H^+ ion), all that remains is the proton in the nucleus. In fact, the H^+ ion is often simply called a proton.

In the internal environment of organisms, the most abundant molecule is water, which carries two pairs of electrons that are not involved in bonding:

$$\ddot{\underset{\cdot\cdot}{O}}\text{ with two H atoms (H—O—H)}$$

A proton (H^+) is strongly attracted to one of these pairs of electrons in a water molecule, thus forming a *hydronium ion* (OH_3^+).

$$H^+ + H_2\ddot{O}: \rightleftarrows \left[H-\ddot{O}-H \atop \quad | \atop \quad H\right]^+$$

In fact, this proton attraction to H_2O is so strong that *no free* H^+ ions exist in aqueous solution: the H^+ ion is always attached to a water molecule. For convenience, however, we will continue to represent the hydrogen ion as H^+.

H^+ concentration and pH

At any given time, a small proportion of the molecules in pure water are ionized to produce hydrogen ions (H+) and hydroxide (OH^-) ions:

$$H_2O \rightleftarrows H^+ + OH^-$$

This reaction is reversible, as indicated by the arrows pointing in both directions. We say that this reaction is at *equilibrium* because in pure water, the rate of ionization of water molecules

$$H_2O \rightarrow H^+ + OH^-$$

is equal to the rate of reaction among the ions.

$$H_2O \leftarrow H^+ + OH^-$$

Thus, at equilibrium there is no further *net* change in concentration, even though both reactions occur continually.

In pure water, this ionization occurs to such an extent that

$$[H^+] = 10^{-7}\ M \text{ and } [OH^-] = 10^{-7}\ M$$

(Square brackets are conventionally used to denote concentrations.) Of course, the concentration of H^+ ions in pure water must equal the concentration of OH^- ions because neither can be formed without the other.

A convenient way of designating the H^+ concentration in a solution is provided by the *pH scale*, which is expressed as the negative logarithms of hydrogen ion concentrations. A *logarithm* is the exponent that indicates the power to which one number must be raised to obtain another number. (For example, the logarithm of 100 to the base 10 is 2.) Thus,

$$pH = \log_{10} \frac{1}{[H^+]}$$

$$(\text{or, } pH = -\log_{10} [H^+])$$

As an example, if $[H^+] = 10^{-3}$ M (.001 M) then pH = 3.

A solution is *neutral* if its H^+ and OH^- concentrations are equal. For water, $[H^+] = [OH^-]$ when both exist in concentrations of 10^{-7} M; that is, when the pH is 7. A solution is *acidic* if it has a higher H^+ concentration than OH^- concentration, and *basic* if its H^+ concentration is lower than its OH^- concentration. Therefore, acidic solutions have pH values below 7, and basic solutions have pH values above 7.

Acids and bases

Substances that change the concentrations of H^+ or OH^- ions when they dissolve in water are known as *acids* and *bases*. An *acid* is a *proton* (*hydrogen ion*) *donor*; it causes an increase in H^+ ion concentration when it ionizes in solution. For example, hydrochloric acid (HCl) is ionized quite completely by the reaction:

$$HCl \rightarrow H^+ + Cl^-$$

Adding hydrochloric acid to a solution increases the relative number of H^+ ions and, therefore, lowers the pH of the solution. The solution of gastric juices in the human stomach, for example, contains a great deal of hydrochloric acid and sometimes may have a pH as low as 1.0. This is such an acidic solution that stomach contents coming in contact with skin could cause a serious acid burn.

A *base* is a *proton acceptor* and has the opposite effect on a solution. For example, sodium hydroxide (NaOH) ionizes almost completely in solution and strongly increases the relative concentration of OH^- ions. Hydroxide ions (OH^-) combine with H+ ions to form water, and thereby decrease the H^+ ion concentration (thus raising the pH of the solution).

Buffers

Even small pH changes may have great effects on biological processes. Because many processes are sensitive to pH changes, it is important that the pH of body fluids remains stable. Cell interiors have an average pH of 7.0 to 7.3; and in humans, for example, blood and tissue fluids have a pH of 7.4 to 7.5. These values remain very stable despite the fact that many biochemical reactions either release or incorporate H^+ ions.

How does pH remain so constant in living things? Such pH stability is possible because organisms have built-in mechanisms to prevent pH change. The most important of these mechanisms are *buffers*. A buffer resists pH change by removing H^+ ions when the H^+ ion concentration rises, and releasing H^+ ions when the H^+ ion concentration falls.

An example of a buffer is the carbonic acid-bicarbonate ion buffer system, which is involved in the buffering of human blood:

$$\underset{\text{Carbonic acid}}{H_2CO_3} \rightleftarrows H^+ + \underset{\text{Bicarbonate ion}}{HCO_3^-}$$

In this example, if H^+ ions are added to the system, they combine with HCO_3^- to form H_2CO_3. This reaction removes extra H^+ ions and keeps the pH from changing. If H^+ ions are removed from the system (for example, if OH^- ions are added and H_2O is formed), more H_2CO_3 will ionize and replace the H^+ ions that were used. Again, pH stability is maintained. This and other buffer systems prevent potentially harmful pH changes in living things.

Organic Molecules

Most of the thousands of kinds of chemical compounds in living organisms are *organic compounds*, compounds that contain carbon. The

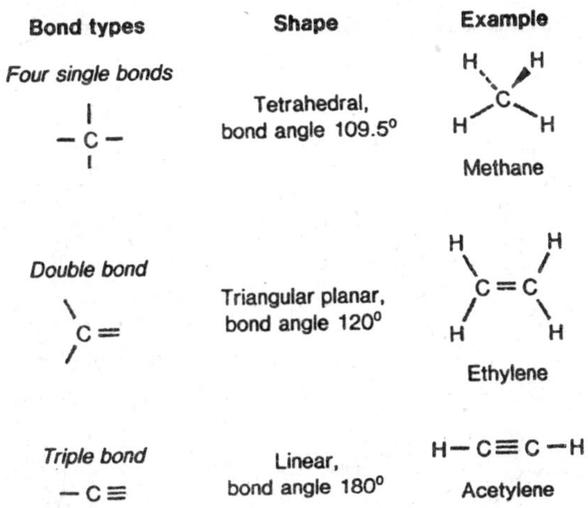

Fig. 3.12. Possible arrangements of bonds around the carbon atom.

most important characteristics of organic compounds depend on properties of this key element, carbon.

The carbon atom has an atomic number of six, with six protons and six electrons. Since such an atom has two electrons in the first shell and four electrons in the second shell, it can form covalent bonds with as many as four other atoms. A carbon atom can bond to a variety of elements, but it most commonly bonds to hydrogen, oxygen, nitrogen, and carbon.

The carbon-to-carbon bonding ability of carbon atoms makes possible carbon chains (sometimes called carbon skeletons) of various lengths and shapes. The carbon skeleton establishes the overall framework of an organic molecule. Carbon atoms can form single, double, or triple covalent bonds with one another, and such bonding patterns determine the characteristics of organic compounds.

Many organic compounds consist of just carbon and hydrogen, but many others contain *functional groups* as well. Functional groups are characteristic patterns of atoms other than those involved in carbon-carbon and carbon-hydrogen bonds. These functional groups are the parts of organic molecules that are most often involved in chemical reactions.

One reason there are so many organic compounds is that carbon forms chains of atoms better than any other element. An unusually long noncarbon chain might contain ten atoms. On the other hand, organic compounds with chains of more than fifty carbon-carbon bonded atoms can be found in living systems, and chains containing carbon

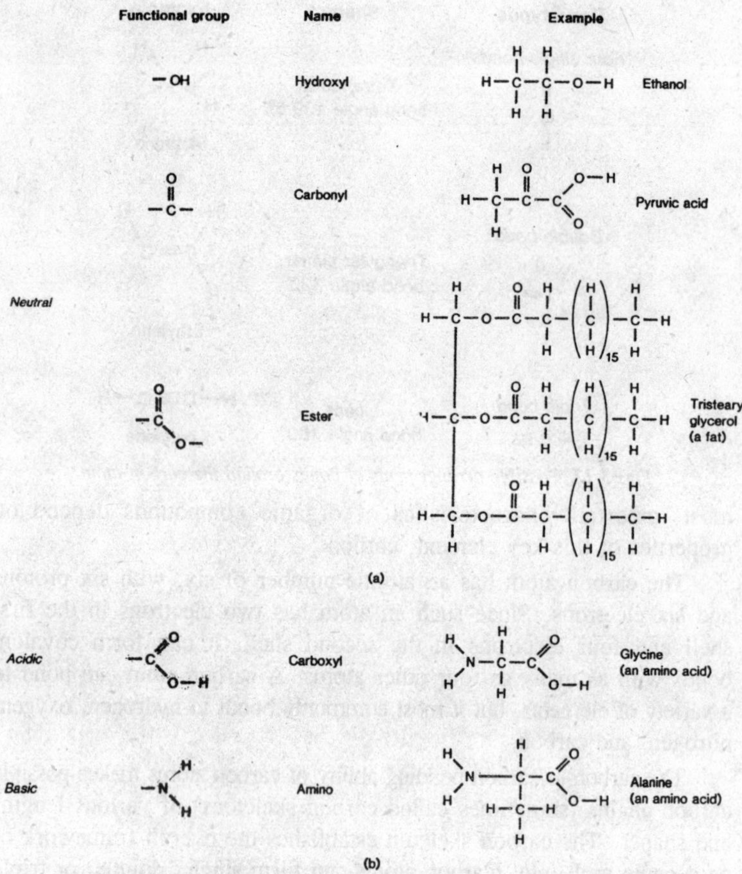

Fig. 3.13. Some examples of functional groups found in organic molecules. (a) Neutral functional groups. (b) Acidic and basic functional groups.

along with other elements such as nitrogen and oxygen are sometimes thousands of atoms long.

Another factor contributing to the diversity of organic compounds is the presence of numerous *isomers*. Isomers are compounds having the same molecular formulas but different atom arrangements. *Structural isomers* are completely different chemical compounds; they share the same molecular formula but have different pairs of atoms connected by bonds. *Geometric isomers* differ from each other only in the orientations of bonds between particular pairs of atoms. These differences are possible in the presence of certain rings or C=C double bonds. Although rotation around C—C single bonds occurs readily,

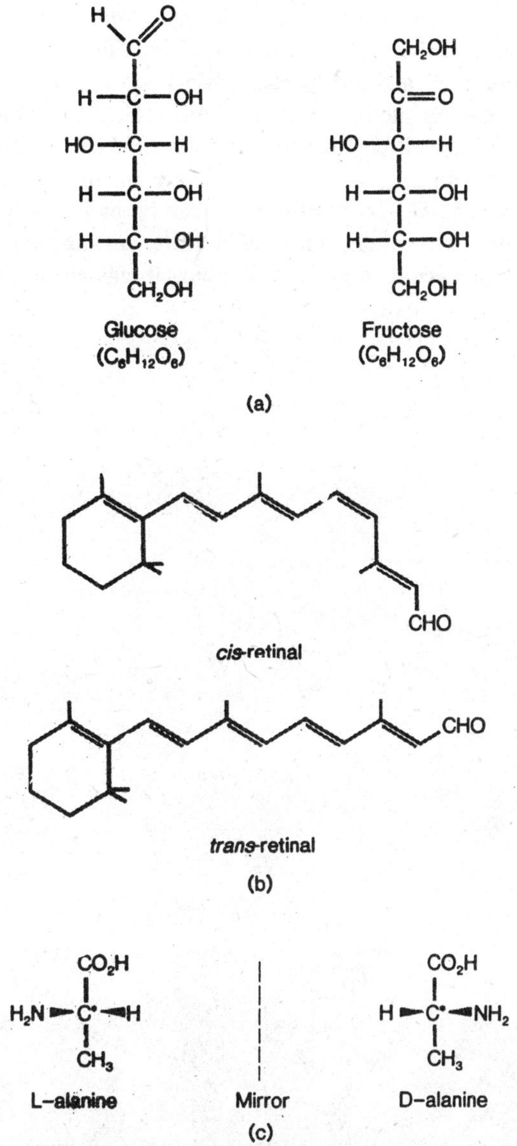

Glucose ($C_6H_{12}O_6$)

Fructose ($C_6H_{12}O_6$)

(a)

cis-retinal

trans-retinal

(b)

L-alanine

Mirror

D-alanine

(c)

Fig. 3.14. Some examples of isomers. (a) Structural isomers are molecules with identical molecular formula, but different pairs of atoms connected by bonds. (b) Geometric isomers are molecules that have different orientations of bonds in space, but that are not minor images of each other. (c) Optical isomers are pairs of molecules that are nonsuperimposable mirror images of each other.

rotation around C=C double bonds does not. *Optical isomers* occur as pairs of nonsuperimposible mirror images, similar to the relationship between your right and left hands. Optical isomers have at least one asymmetric carbon atom, a carbon atom attached to four different groups. This makes it possible for the four different groups to occupy two different arrangements in space. These mirror-image forms are designated L- and D-. Such differences can be biologically important. For example, it is interesting and a little curious that all of the amino acids that occur in proteins in living things are of the L- form.

4

MOLECULES IN EARLIEST CELLS

When *On the Origin of Species* appeared in 1859, the history of life could be traced back to the beginning of the Cambrian period of geologic time, to the earliest recognized fossils, forms .hat are now known to have lived more than 500 million years ago. A far longer prehistory of life has since been discovered: it extends back through geologic time almost three billion years more. During most of that long Precambrian interval the only inhabitants of the earth were simple microscopic organisms, many of them comparable in size and complexity to modern bacteria. The conditions under which these organisms lived differed greatly from those prevailing today, but the mechanisms of evolution were the same. Genetic variations made some individuals better fitted than others to survive and to reproduce in a given environment, and so the heritable traits of the better-adapted organisms were more often represented in succeeding generations. The emergence of new forms of life through this principle of natural selection worked great changes in turn on the physical environment, thereby altering the conditions of evolution.

One momentous event in *Precambrian* evolution was the development of the biochemical apparatus of oxygen-generating photosynthesis. Oxygen released as a by-product of photosynthesis accumulate in the atmosphere and effected a new cycle of biological adaptation. The first organisms to evolve in response to this environmental change could merely tolerate oxygen; later cells could actively employ oxygen in metabolism and were thereby enabled to extract more energy from foodstuff.

A second important episode in Precambrian history led to the emergence of a new kind of cell, in which the genetic material is aggregated in a distinct nucleus and is bounded by a membrane. Such nucleated cells are more highly organized than those without nuclei. What is most important, only nucleated cells are capable of advanced sexual reproduction, the process whereby the genetic variations of the parents can be passed on to the off spring in new combinations. Because sexual reproduction allows novel adaptations to spread quickly through a populationists development accelerated the pace of evolutionary change. The large, complex, multicellular forms of life that have appeared and quickly diversified since the beginning of the *Cambrian period* are without exception made up of nucleated cells.

The history of life in its later phases, since the start of the *Cambrian period*, has been reconstructed mainly from the study of fossils preserved in sedimentary rocks. In the 18th and 19th centuries it gradually became apparent that the fossil record has appreciable chronological and geographical continuity. The fossil deposits form recognizable layers which can be identified in widely separated geological formations. Boundaries between such layers, where one characteristic suite of fossils gives way to another, provide the basis for dividing geologic time into eras, periods and epochs.

One of the most dramatic boundaries in the rock record is the one that separates the Cambrian period from all that came before. The 11 periods of geologic time since the start of the Cambrian are referred to collectively as the *Phanerozoic era*, which might be translated from the Greek as the era of manifest life. The preceding era is called simply the Precambrian.

By itself the geologic time scale cannot provide dates for fossil deposits; it only lists their sequence. Ages can be calculated, however, from the constant rate of decay of radioactive isotopes in the earth's crust. By determining how much of an isotope has decayed since the minerals in a rock unit crystallized, a date can be assigned to that unit and to nearby strata containing fossils. Radioactive-isotope studies of this kind, carried out on rocks from many parts of the world, have established a rather well-defined date for the start of the Phanerozoic era: it began about 570 million years ago. The same method indicates that the earth itself and the rest of the solar system are 4.6 billion years old. Thus the Precambrian era encompasses some seven-eighths of the earth's entire history.

The boundary between the Precambrian era and the Cambrian period has traditionally been viewed as a sharp discontinuity. In Cambrian strata there are abundant fossils of marine plants and animals:

seaweeds, worms, sponges, mollusks, lampshells and, what are perhaps most characteristic of the period, the early arthropods called trilobites. It was thought from many years that fossils were entirely absent in the underlying Precambrian strata. The Cambrian fauna seemed to come into existence abruptly and without known predecessors.

Life could not have begun with organisms as complex as *trilobites*. In *On the Origin of Species* Darwin wrote: "To the question why we do not find rich fossiliferous deposits belonging to....periods prior to the Cambrian system. I can give no satisfactory answer... The case at present must remain inexplicable; and may be truly urged as a valid argument against the views here entertained." The argument is no longer valid, but it is only in the past 20 years or so that a definitive answer to it has been found.

One part of the answer lies in the discovery of primitive fossil animals in rocks below the earliest Cambrian strata. The fossils include the remains of jellyfishes, various kinds of worms and possibly sponges, and they make up a fauna quite distinct from that of the predominantly shelled animals of the Cambrian period. These discoveries however, extend the fossil record by only about 100 million years, less than four per cent of the Precambrian era. It can still be asked: What came before?

Since the 1950's a far-reaching explanation has emerged. It has come to be recognized that not only are many Precambrian rocks fossil-bearing but also Precambrian fossils can be found even in some of the most ancient sedimentary deposits known. These fossils had escaped notice earlier largely because they are the remains only of microscopic forms of life.

An important clue in the search for Precambrian life was discovered in the early years of the 20th century, but its significance was not fully appreciated until much later. The clue came in the form of masses of thinly layered limestone rock discovered by *Charles Doolittle Walcott* in Precambrian strata from western North America. Walcott found numerous moundlike or pillar like structures made up of many draped horizontal layers, like, tall stacks of pancakes. These structures are now called stromatolites, from the Greek *stroma* meaning bed or coverlet, and *lithos*, meaning stone.

Walcott interpreted the stromatolites as being fossilized reefs that had probably been formed by various types of algae. Other workers were skeptical, and for many years the stromatolites were widely attributed to some non biological origin. The first convincing evidence substantiating Walcott's hypothesis came in 1954, when *Stanley A. Tyler* of the University of Wisconsin and *Elso S. Barghoorn* of Harvard University reported the discovery of fossil microscopic plants in an outcropping of Precambrian rocks called the Gunflint Iron formation near Lake Superior in Ontario. Most of the Gunflint fossils, which form the layers of dome-shaped and pillar-like stromatolites, resemble modern blue-green algae and bacteria. More recently, living stromatolites have been identified in several coastal habitats, most notably in a lagoon at Shark Bay on the western coast of Australia. They are indeed built up by communities of blue-green algae and bacteria, and they are strikingly similar in form to the fossilized Precambrian structures.

Today microfossils have been identified in some 45 stromatolitic deposits (All but three of these fossilized communities have been found in the past 10 years). The fossils are often well preserved, the cell walls being petrified in three-dimensional form, and they have become a prime source of documentation for the early history of life. In recent years the search for Precambrian microfossils in other kinds of sediments, such as shales deposited in offshore environments, has also been rewarded. These fossils are generally not as well preserved as the ones in stromatolites, most of them having been flattened by pressure; on the other hand, they supply information about Precambrian life in a habitat quite different from that of the shallow-water stromatolites.

A surprising amount of information can be derived from the fossil remains of a microorganism. Size, shape and degree of morphological complexity are among the most easily recognized features, but under favourable circumstances even details of the internal structure of cells can be discerned. In retracing the course of Precambrian evolution, however, there is no need to rely exclusively on the fossil record. An entirely independent archive has been preserved in the metabolism and the biochemical pathways of modern, living cells. No living organism is biochemically identical with its Precambrian antecedents, but vestiges of earlier biochemistries have been retained. By studying their distribution in modern forms of life it is sometimes possible to deduce when certain biochemical capabilities first appeared in the evolutionary sequence.

Still another independent source of information about the early evolutionary progression is based neither on living nor on fossil organisms but on the inorganic geological record. The nature of the minerals found there reflects physical conditions at the time the minerals were deposited, conditions that may have been influenced by biological innovations. In order to understand the introduction of oxygen into the early atmosphere, for example, all three fields of study must be called on to testify, the mineral record tells when the change took place the fossil record reveals the organisms responsible and the distribution of biochemical capabilities among modern organisms puts the development in its proper evolutionary context.

Since the 1960's it has become apparent that the greatest division among living organisms is not between plants and animals but between organisms whose cells have nuclei and those that lack nuclei. In terms of biochemistry, metabolism, genetics and intracellular organization, plants and animals are very similar; all such higher organisms, however, are quite different in these features from bacteria and blue-green algae, the principal types of non-nucleated life. Recognition of this discontinuity has been important for understanding the early stages of biological history.

Organisms whose cells have nuclei are called eukaryotes, from the Greek roots *eu-*, meaning well or true, and *karyon,* meaning kernel or nut. Cells without nuclei are prokaryotes, the prefix promeaning before. All green plants and all animals are eukaryotes. So are the fungi, including the molds and the yeasts, and protists such as *Paramecium* and *Euglena.* The prokaryotes include only two groups of organisms, the bacteria and the blue-green algae. The latter produce oxygen through photosynthesis like other algae and higher plants, but they have much stronger affinities with the bacteria than they do with eukaryotic forms of life. Later on several workers refer to blue-green algae by an alternative and more descriptive name, the *cyanobacteria.*

Several important traits distinguish eukaryotes from prokaryotes. In the nucleus of a eukaryotic cell the DNA is organized in chromosomes and is enclosed by an intracellular membrane, many prokaryotes have only a single loop of DNA which is loose in the cytoplasm of the cell. Prokaryotes reproduce asexually by the comparatively simple process of binary fission. In contrast, asexual reproduction in eukaryotic cells takes place through the complicated process of mitosis, and most eukaryotes can also reproduce sexually through meiosis and the subsequent fusion of sex cells. (The "parasexual" reproduction of some prokaryotes differs markedly from advanced eukaryotic sexuality). Eukaryotic cells are generally larger than prokaryotic ones, although the range of sizes overlaps, and almost all

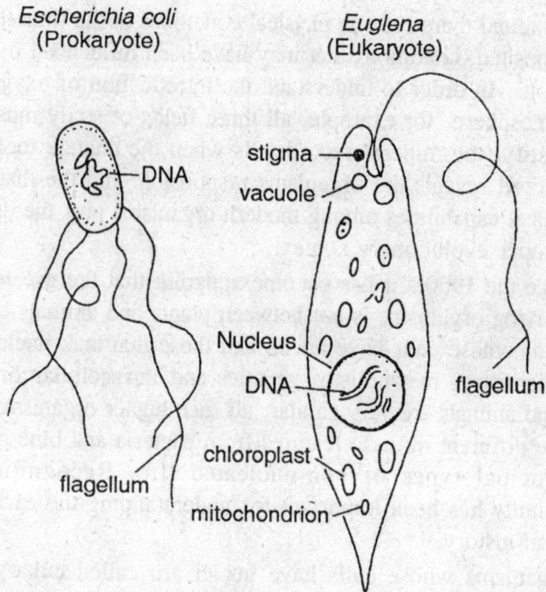

Fig. 4.1. Greatest division among organisms in the one separating cells with nuclei (eukaryotes) from those without nuclei (prokaryotes).

prokaryotes are unicellular organisms whereas the majority of eukaryotes are large, complex and many celled. A mammalian animal, for example, can be made up of billions of cells, which are highly differentiated in both structure and function.

An intriguing feature of eukaryotic cells is that they have within them smaller membrane-bounded subunits, or organelles, the most notable being mitochondria and chloroplasts. Mitochondra are present in all eukaryotes, where they play a central role in the energy economy of the cell. Chloroplasts are present in some protists and in all green plants and are responsible for the photosynthetic activities of those organisms. It has been suggested that both mitochondria and chloroplasts may be evolutionary derivatives of what were once free-living microorganisms, an idea discussed in particular by Lynn Margulis of Boston University. The modern chloroplast, for example, may be derived from a cyanobacterium that was engulfed by another cell and that later established a symbiotic relationship with it. In support of this hypothesis it has been noted that both mitochondria and chloroplasts contain a small fragment of DNA whose organization is somewhat like that of prokaryotic DNA. In the past several years the testing of this hypothesis has generated a large body of data on the comparative biochemistry of modern microorganism, data that also provide clues to the evolution of life in the Precambrian.

One further difference between prokaryotes and eukaryotes is of particular importance in the study of their evolution: the extent to which the two types of organisms tolerate oxygen. Among the prokaryotes oxygen requirements are quite variable. Some bacteria cannot grow or reproduce in the presence of oxygen: they are classified as obligate anaerobes. Others can tolerate oxygen but can also survive in its absence; they are facultative anaerobes. There are also prokaryotes that grow best in the presence of oxygen but only at low concentrations, far below that of the present atmosphere. Finally, there are fully aerobic prokaryotes forms that cannot survive without oxygen.

In contrast to this variety of adaptations the eukaryotes present a pattern of great consistency: with very few exceptions they have an absolute requirement for oxygen, and even the exceptions seem to be evolutionary derivatives of oxygen-dependent organisms. This observation leads to a simple hypothesis: the prokaryotes evolved during a period when environmental oxygen concentrations were changing, but by the time the eukaryotes arose the oxygen content was stable and relatively high.

One indication that eukaryotic cells have always been aerobic is provided by mitotic cell division, a process that can be considered a definitive characteristic of the group. Many eukaryotic cells can survive temporary deprivation of oxygen and can even carry on some metabolic functions; it appears that no cell, however, can undergo mitosis unless oxygen is available at least in low concentration .

The pathways of metabolism itself—the biochemical mechanisms by which an organism extracts energy from foodstuff—provide more detailed evidence. In eukaryotes the central metabolic process is respiration, which in overall terms can be described as the burning of the sugar glucose with oxygen to yield carbon dioxide, water and energy. Some prokaryotes (the aerobic or facultative ones) are also capable of respiration, but many derive their energy solely from the simpler process of fermentation. In bacterial fermentation glucose is not combined with oxygen (or with any othersubstance from outside the cell) but is simply broken down into smaller molecules. In both respiration and fermentation part of the energy released through the decomposition of glucose is captured in the form of high-energy phosphate bonds, usually in molecules of adenosine triphosphate (ATP). The rest of the energy is lost from the cell as heat.

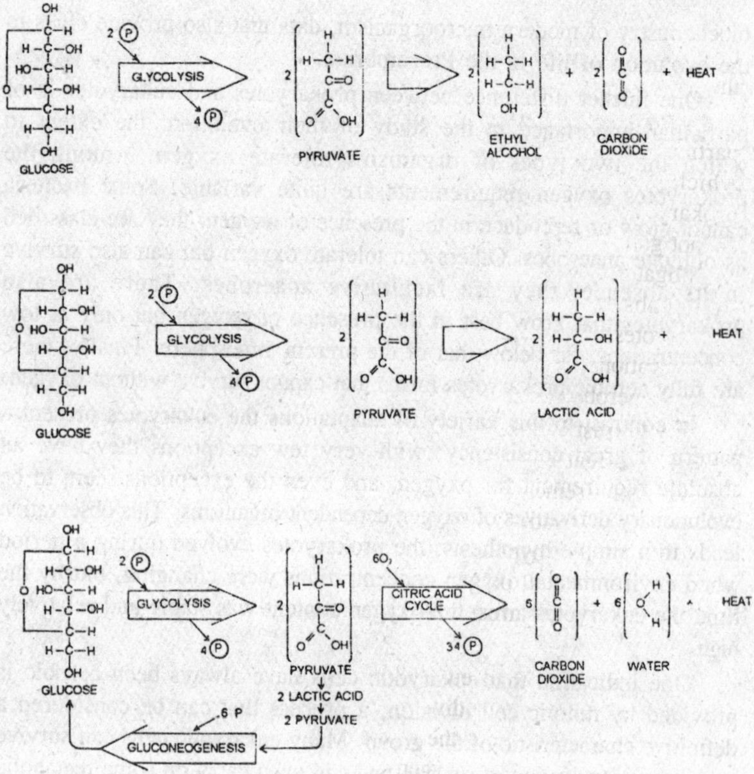

Fig. 4.2. Metabolic pathways by which cells extract energy from foodstuff apparently evolved in response to an increase in free oxygen.

Respiratory metabolism has two main components: a short series of chemical reactions, collectively called glycolysis, and a longer series called the citric acid cycle. In glycolysis a glucose molecule, with six carbon atoms, is broken down into two molecules of pyruvate, each having three carbon atoms. No oxygen is required for glycolysis, but on the other hand it releases only a little energy with a net gain of only two molecules of ATP.

The fuel for the citric acid cycle is the pyruvate formed by glycolysis. Through a series of enzyme-controlled reactions the carbon atoms of the pyruvate are oxidized and the oxidations are coupled to other reactions that result in the synthesis of ATP. For each two molecules of pyruvate (and hence for each molecule of glucose entering the sequence) 34 additional molecules of ATP are formed. The complete respiratory pathway is thus far more effective than glycolysis alone.

In respiration the proportion of energy released that can be recovered in useful form (as ATP) is higher than it is in fermentation, about 38 per cent instead of only some 30 per cent, and in respiration the net energy yield to the cell is some 18 times greater. By breaking down the glucose to simple inorganic molecules (carbon dioxide and water) respiration liberates virtually all the biologically usable energy stored in the chemical bonds of the sugar.

The metabolism of the prokaryotes immediately suggests an evolutionary relationship between them and the eukaryotes; up to a point fermentation is indistinguishable from glycolysis. In bacterial fermentation a molecule of glucose is split into two molecules of pyruvate, with a net yield of two molecules of ATP. As in glycolysis, no oxygen is required for the process. In anaerobic prokaryotes, however, the metabolic pathway essentially ends at pyruvate. The only further reactions transform the pyruvate into such compounds as lactic acid, ethyl alcohol or carbon dioxide, which are excreted by the cell as wastes.

The similarity of fermentation in prokaryotes to glycolysis in eukaryotes seems too close to be a coincidence, and the assumption of an evolutionary relationship between the two groups provides a ready explanation. It seems likely that anaerobic fermentation became established as an energy-yielding process early in the history of life. When atmospheric oxygen became available for metabolism, it offered the potential for extracting 18 times as much useful energy from carbohydrate: a net yield of 36 molecules of ATP instead of only two molecules. The oxygen-dependent reactions did not however, simply replace the anaerobic ones: they were appended to the existing anaerobic pathway.

Further evidence for this proposed evolutionary sequence can be found in the behaviour of some eukaryotic cells under conditions of oxygen deprivation. In mammalian muscle cells, for example, prolonged exertion can demand more oxygen than the lungs and the blood can supply. The citric acid cycle is then disabled, but the cells continue to function, albeit at reduced efficiency, through glycolysis alone. Under such conditions of oxygen debt pyruvate is not consumed in the cell, but in the liver it can be converted back into glucose (at a cost in energy of six ATP molecules). Significantly the pyruvate itself is not transported to the liver but instead is converted into lactic acid, which in the liver must then be returned to the form of pyruvate. This use of lactic acid may represent a vestige of an earlier, bacterial pathway that under aerobic conditions has been suppressed. Indeed, the oxygen-starved muscle cell seems to revert to a more primitive, entirely anaerobic form of metabolism.

The development of an oxygen-dependent biochemistry can also be traced through a consideration of reaction sequences in the synthesis of various biological molecules. Once again stages in the synthetic pathway that emerged early in the Precambrian can be expected to proceed in the absence of oxygen. Reaction steps nearer the end product of the pathway, which were presumably added at a later age, might with increasing frequency require oxygen. The distribution of the oxygen-demanding steps among various kinds of organisms could also have evolutionary significance. If only one pathway has evolved for the synthesis of a class of biochemical substances, then primitive forms of life might be expected to exhibit only the initial, anaerobic steps. Organisms that arose later might exhibit progressively longer, oxygen-dependent synthetic sequences.

In aerobic organisms it might seem at first that virtually all biochemical synthesis require oxygen; eukaryotic cells exhibit relatively little synthetic activity under anoxic conditions. For the most part, however, the oxygen requirement of such synthesis is simply for metabolism: the construction of biological molecules demands energy in the form of ATP, and most of the ATP is supplied through the oxygen-dependent citric acid cycle. If ATP is made available from some other source many synthetic pathways can proceed unimpaired.

Some synthesis, however have an intrinsic requirement for oxygen, quite apart from metabolic demands. Molecular oxygen is needed, for example, in the synthesis of bile pigments in vertebrates, of chlorophyll *a* in higher plants and of the amino acids hydroxyproline and, in animals tyrosine. The oxygen dependence of two synthetic pathways in particular has been determined in detail. One of these pathways controls the manufacture of a class of compounds that includes the sterols and the carotenoids and the other is concerned with the synthesis of fatty acids.

Sterols, such as cholesterol and the steroid hormones, are flat, platelike molecules derived from the compound squalene, which has 30 carbon atoms. Carotenoids are derived from the 40-carbon compound phytoene: they are pigments, such as carotene, the orange-yellow compound in carrots, and they are found in virtually all photosynthetic organisms. A common starting point for the synthesis of both groups of compounds is isoprene, a five-carbon molecule that is also the repeating unit in synthetic rubber. In the biological synthesis two isoprene subunits are joined head to tail then a third isoprene is added to form a 15-carbon polymer, farnesyl pyrophosphate. At this point there is a fork in the pathway. In one continuaion of the synthesis two farnesyl chains are joined to form squalene, the 30-carbon precursor of the sterols. In the other continuation a fourth isoprene subunit is added, and only then are two of the chains joined. The product in this case is phytoene, the 40-carbon precursor of the carotenoids and of other pigments derived from them, such as the xanthophylls.

Up to this step in the synthetic pathway none of the reactions requires the participation of molecular oxygen. The next step in the synthesis of sterols, however, is the conversion of the linear squalene molecule to a 30-carbon ring, and this transformation does require oxygen; so do most of the subsequent steps in sterol synthesis. On the other branch of the pathway there are a few more anaerobic reactions, and indeed carotenoids can be made from phytoene without oxygen. Several further modifications of the carotenoids, however, such as the production of the pigments called epoxy-xanthophylls, are oxygen dependent.

Two observations about the evolution of these biosynthetic pathways are appropriate. Even in groups of organisms that have long been aerobic the first steps in the synthesis are independent of the oxygen supply; molecular oxygen enters the reaction sequence only at later stages. In a similar way the most primitive living organisms, the anaerobic bacteria, are capable only of the first segments of the pathway, the anaerobic segments. The more complex aerobic bacteria and the photosynthetic cyanobacteria have longer synthetic pathways, including some steps that require oxygen. Advanced eukaryotes, such as vertebrate animals and higher plants, have long, branched synthetic pathways, with many steps in which molecular oxygen is required.

A similar pattern can be discerned in the synthesis of fatty acids and their derivatives. The fatty acids are straight carbon-chain compounds that have a carboxyl group (COOH) at one end. A fatty acid is said to be saturated if there are no double bonds between carbon atoms in the chain: it is saturated with hydrogen, which fills all the available bonding positions. An unsaturated fatty acid has a double bond between two carbon atoms or it may have several such double bonds: for each double bond two hydrogen atoms must be removed from the molecule.

In the synthesis of fatty acids the molecule grows by the repeated addition of units two carbon atoms long. The first few steps in the synthesis are identical in all organisms, and they yield fully saturated fatty acids. The first branch in the pathway comes when the developing chain is eight carbons long. At that point many prokaryotes can introduce a double bond, which eukaryotes cannot. There is a second branch at the next step when the saturated chain is 10 carbons long: a double bond can similarly be introduced at that point by many prokaryotes but not by eukaryotes. No matter which branch is followed, elongation of the chain ends at 18 carbons. At that point the fatty acids produced by many prokaryotes contain a double bond, but in eukaryotes the product is always the fully saturated molecule, stearic acid. None of the steps in this sequence, whether in prokaryotes or eukaryotes, requires molecular oxygen.

If no subsequent transformations of fatty acids were possible, eukaryotic cells would be incapable of synthesizing any but the fully saturated forms. Actually extensive modifications can be accomplished through the process of oxidative desaturation, in which double bonds are formed by removing two hydrogen atoms and combining them with oxygen to form water, Oxidative desaturation can take place only in the presence of molecular oxygen (O_2). Through this mechanism cyanobacteria make unsaturated fatty acids with two, three and four double bonds, and eukaryotes form polyunsaturated fatty acids (with multiple double bonds).

As in the sterol-carotenoid synthesis, an analysis of the fatty-acid pathway argues for a pattern of biochemical evolution in which the increasing availability of atmospheric oxygen played a central role. The first steps in the synthetic sequence are common to all organisms capable of making fatty acids, and in the most primitive organisms those are the only steps. Hence the reactions that come first in the biochemical sequence apparently also developed early in the history of life; these first steps are all anaerobic. Organisms that presumably emerged somewhat later (such as aerobic bacteria and cyanobacteria) have longer pathways, including a few steps of oxidative desaturation. In advanced eukaryotes a substantial proportion of the steps are oxygen dependent.

Comparisons of the metabolism and biochemistry of prokaryotes and eukaryotes thus provide strong evidence that the latter group arose only after a substantial quantity of oxygen had accumulated in the atmosphere. Hence it is of interest to ask when eukaryotic cells first appeared. It seems apparent that an oxygen-rich atmosphere cannot have developed later than this signal evolutionary event.

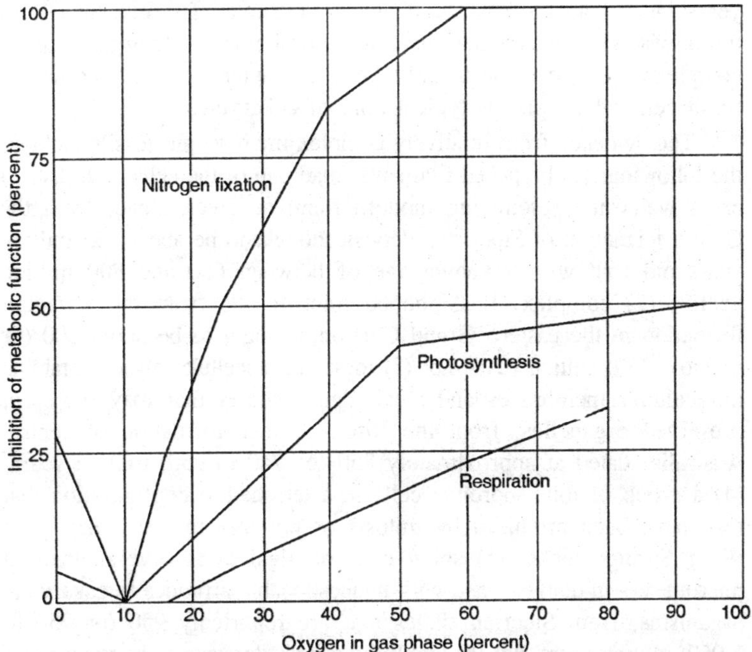

Fig. 4.3. Oxygen inhibition of metabolic functions in cyanobacteria suggests that these aerobic prokaryotes are adapted to an optimum oxygen concentration of about 10 percent, or roughly half the oxygen concentration of the earth's present atmosphere.

The primary means of assigning a date to the origin of the eukaryotes is through the fossil record. Because this field of study is so new, however, the available information is scanty and often difficult to interpret. It is rarely a straightforward task to identify a microscopic, single-cell organism as being eukaryotic merely from an examination of its fossilized remains. And even when a fossil has been identified as unequivocally eukaryotic the available radioactive-isotope methods of dating can rarely assign it a precise age. At best such methods have an accuracy of only about *plus* or minus 5 per cent. What is more, the age determinations are generally carried out on rocks that were once molten, such as volcanic lavas, whereas the fossils are found in sedimentary deposits. Consequently the stratum of the fossil itself usually cannot be dated: it is merely assigned an age somewhere between the ages of the nearest underlying and overlying datable rock units.

In spite of these difficulties there is now substantial evidence for the existence of eukaryotic fossils in rocks hundreds of millions of years older than the earliest Phanerozoic strata. The evidence is of two kinds: microfossils that display a morphological or organizational complexity judged to be of eukaryotic character and the presence of fossil cells whose size is typical only of eukaryotes.

The evidence from relatively complex microscopic fossils includes the following: (1) branched filaments, made up of the cells with distinct cross walls and resembling modern fungi or green algae, from the Olkhin formation of Siberia, a deposit thought to be about 725 million years old (but with a known age of between 680 and 800 million years); (2) complex, flask-shaped microfossils from the Kwagunt formation in the eastern Grand Canyon, thought to be about 800 (or 650 to 1,150) million years old; (3) fossils of unicellular algae containing intracellular membranes and small dense bodies that may represent preserved organelles, from the Bitter Springs formation of central Australia, dated at approximately 850 (or 740 to 950) million years; (4) a group of four sporelike cells in a tetrahedral configuration that may have been produced by mitosis or possibly meiosis, also from Bitter Springs rocks; (5) spiny cells or algal cysts several hundred micrometers in diameter and with unquestionable affinities to eukaryotic organisms, from Siberian shales that are reportedly 950 (or 750 to 1,050) million years old; (6) highly branched filaments of large diameter and with rare cross walls, similar in some respects to certain green or golden-green eukaryotic algae, from the Beck Spring dolomite of southeastern California (1,300, or 1,200 to 1,400 million years old) and from the Skillogalee dolomite of South Australia (850, or 740 to 867 million years old); (7) spheroidal microfossils described as exhibiting two-layered walls and having "medial splits" on their surface and which may represent an encystment stage of a eukaryotic alga from shales 1,400 (or 1,280 to 1,450) million years old in the McMinn formation of northern Australia; (8) a tetrachedral group of four small cells, resembling spores produced by mitotic cell division of some green algae, from the Amelia dolomite of northern Australia, approaching 1,500 (or 1,390 to 1,575) million years in age; (9) unicellular fossils that appear to be exceptionally well preserved and that are reported to contain small membrane-bounded structures that could be remnants of organelles, from the Bungle-Bungle dolomite in the same region as the Amelia dolomite and of approximately the same age.

Thus the earliest of these eukaryote-like fossils are probably somewhat less than 1,500 million years old. Numerous types of

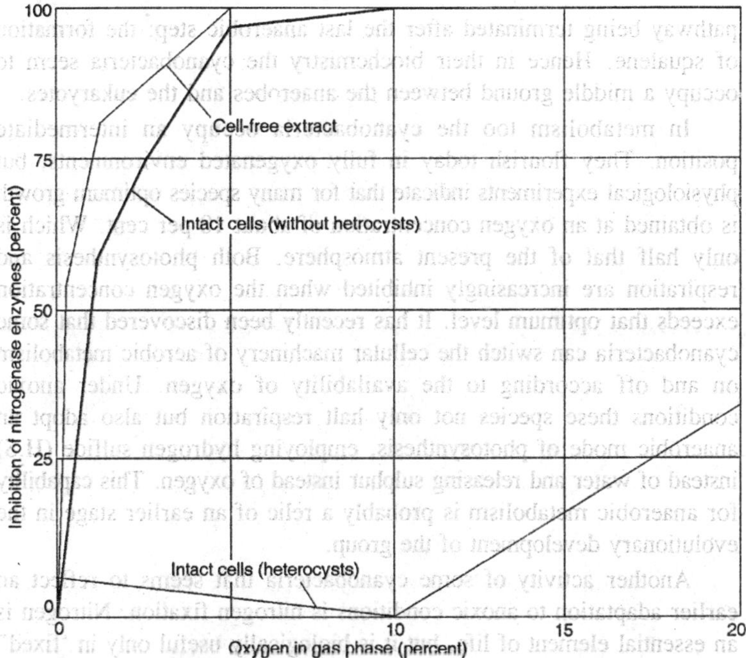

Fig. 4.4. Inhibition of nitrogen fixation in the presence of oxygen is caused by the deactivation of the nitrogenase enzymes.

microfossils have been discovered in older sediments, but none of them seems to be a strong candidate for that in overall effect (although not in mechanism) is the reverse of respiration. The energy of sunlight is employed to make carbohydrates from water and carbon dioxide, and molecular oxygen is released as a by-product. The cyanobacteria can tolerate the oxygen they produce and can make use of it both metabolically (in aerobic respiration) and in synthetic pathways that seem to be oxygen dependent (as in the synthesis of chlorophyll *a*). Nevertheless, the biochemistry of the cyanobacteria differs from that of green eukaryotic plants and suggests that the group originated during a time of fluctuating oxygen concentration. For example, although many cyanobacteria can make unsaturated fatty acids by oxidative desaturation some of them can also employ the anaerobic mechanism of adding a double bond during the elongation of the chain. In a similar manner oxygen-dependent synthesis of certain sterols can be carried out be some cyanobacteria, but the amounts of the sterols made in this way are minuscule compared with the amounts typical of eukaryotes. In other cyanobacteria those sterols are not found at all, the biosynthetic pathway being terminated after the last anaerobic step: the formation of squalene. Hence in their biochemistry the cyanobacteria seem to occupy a middle ground between the anaerobes and the eukaryotes.

In metabolism too the cyanobacteria occupy an intermediate position. They flourish today in fully oxygenated environments, but physiological experiments indicate that for many species optimum growth is obtained at an oxygen concentration of about 10 per cent. Which is only half that of the present atmosphere. Both photosynthesis and respiration are increasingly inhibited when the oxygen concentration exceeds that optimum level. It has recently been discovered that some cyanobacteria can switch the cellular machinery of aerobic metabolism on and off according to the availability of oxygen. Under anoxic conditions these species not only halt respiration but also adopt an anaerobic mode of photosynthesis, employing hydrogen sulfide (H_2S) instead of water and releasing sulphur instead of oxygen. This capability for anaerobic metabolism is probably a relic of an earlier stage in the evolutionary development of the group.

Another activity of some cyanobacteria that seems to reflect an earlier adaptation to anoxic conditions is nitrogen fixation. Nitrogen is an essential element of life, but it is biologically useful only in 'fixed" form, for example combined with hydrogen in ammonia (NH_3). Only prokaryotes are capable of fixing nitrogen (although they often do so in symbiotic relationships with higher plants). The crucial complex of enzymes for fixation, the nitrogenases, is highly sensitive to oxygen. In cell-free extracts nitrogenases are partially inhibited by as little as ×1 per cent of free oxygen, and they are irreversibly inactivated in minutes by exposure to oxygen concentrations of only about 5 per cent.

Such a complex of enzymes could have originated only under anoxic conditions, and it can operate today only if it is protected from exposure to the atmosphere. Many nitrogen-fixing bacteria provide at protection simply by adopting an anaerobic habitat, but among the cyanobacteria a different strategy has developed; the nitrogenase enzymes are protected in specialized cells, called heterocysts, whose internal milieu is anoxic. The heterocysts lack certain pigments essential for photosynthesis, and so they generate no oxygen of their own. They have thick cell walls and are surrounded by a mucilaginous envelope that retards the diffusion of oxygen into the cell. Finally, they are equipped with respiratory enzymes that quickly consume any uncombined oxygen that may leak in.

Because of the thick cell walls heterocysts should be comparatively easy to recognize in fossil material. Indeed, possible heterocysts have been reported from several Precambrian rock units, the oldest being about 2.2 billion years in age. If these cells are indeed heterocysts, they may be taken as a sign that free oxygen was present by then, at least in small concentrations.

Nitrogen fixation has a high cost in energy, and the capability for it would therefore seem to confer a selective advantage only when fixed nitrogen is a scarce resource. Today the main sources of fixed nitrogen are biological and industrial, but biologically usable nitrate (NO_3-) is formed by the reaction of atmospheric nitrogen and oxygen. In the anoxic atmosphere of the early Pre-identification as eukaryotic for example, the well-studied Canadian fossils of the Gunflint and Belcher Island iron formations, which are about two billion years old, have been interpreted as exclusively prokaryotic.

The testimony of these as yet rare and unusual specimens can be checked through statistical studies of the sized of known Precambrian microfossils. The size ranges of prokaryotes and eukaryotes overlap, so that a particular fossil cannot always be classified unambiguously on the basis of size alone; by cataloguing the measured sized in a large sample of fossils, however, it may be possible to determine whether or not eukaryotic cells are present. Among modern species of spheroidal cyanobacteria about 60 per cent are very small, less than five micrometers in diameter: of the remaining species only a few are larger than 20 micrometers and none is larger than 60 micrometers. Unicellular eukaryotes, such as green or red algae, can be much larger. Typically they fall in the range between 5 and 60 micrometers, but several per cent of living species are larger than 60 micrometers and a few are larger than 1,000 micrometers (one millimeter).

Systematic size measurements have been made on some 8,000 fossil cells from 18 widely dispersed Precambrian deposits. On the basis of those data certain tentative conclusions can be drawn. Cells larger than 100 micrometers, and hence of distinctly eukaryotic dimensions, are unknown in rocks older than about 1,450 million years. Virtually all the unicellular fossils from rocks of that age, whether they grew in shallow-water stromatolites or were deposited in offshore shales, are of prokaryotic size.

Cells larger than modern prokaryotes (greater than 60 micrometers in diameter) first become abundant in rocks about 1,400 million years old. Algae of this type were apparently free-floating rather than mat-forming species, and they are therefore particularly common in shales, sediments deposited in deeper water. Such eukaryote-size fossils have been known for several years from shales of this age in China and in the U.S.S.R. Recently cells more than 100 micrometers in diameter have also been discovered in the Newland limestone of Montana, and cells more than 600 micrometers in size (10 times the size of the largest spheroidal prokaryote) have been found in the McMinn formation of Australia; the age of both of these fossil-bearing deposits is about 1,400 million years.

In somewhat younger Precambrian sediments there are still larger cells, fossils greater than one millimeter in diameter (with some as large as eight millimetres). They were first described in 1899 by Walcott, who discovered them in rocks from the Grand Canyon. They have since been found in nearly a dozen other rock units throughout the world. The oldest seem to be those from Utah and from Siberia, each about 950 million years old, and those from northern India, which could be even older (from 910 to 1,150 million years old).

Studies of both the morphology and the size of unicellular fossils therefore suggest that there is a break in the fossil record between 1,400 and 1,500 million years ago. Below this horizon cells with eukaryotelike traits are rare or absent; above it they become increasingly common. Moreover, the data suggest that the diversification of the eukaryotes began shortly after the cell type first appeared, apparently within the next few hundred million years. By a billion years ago there had been substantial increases in cell size, in morphological complexity and in the diversity of species. All these indicators also suggest, of course, that oxygen-dependent metabolism, which is highly developed even in the most primitive eukaryotes, had already become established by about 1.5 billion years ago.

The prokaryotes that must have held exclusive sway over the earth before the development of eukaryotic cells were less diverse in form, but they were probably more varied in metabolism and biochemistry than their eukaryotic descendants. Like modern prokaryotes, the ancient species presumably varied over a broad range in their tolerance of oxygen, all the way from complete intolerance to absolute need. In this regard one group of prokaryotes, the cyanobacteria, are of particular interest in that they were largely responsible for the development of an oxygen-rich atmosphere.

Like higher plants cyanobacteria carry out aerobic photosynthesis, a process cambrian the latter mechanism would obviously have been impossible. The lack of atmospheric oxygen would also have indirectly reduced the concentration of ammonia to very low levels. Ammonia is dissociated into nitrogen and hydrogen by ultraviolet radiation, most of which is filtered out today by a layer of ozone (O_3) high in the atmosphere; without free oxygen there would have been little ozone, and without this protective shield atmospheric ammonia would have been quickly destroyed.

It is likely that the capability for nitrogen fixation developed early in the Precambrian among primitive prokaryotic organisms and in an environment where fixed nitrogen was in short supply. The vulnerability of the nitrogenase enzymes to oxidation was of no consequence then, since the atmosphere had little oxygen. Later, as the photosynthetic activities of the cyanobacteria led to an increase in atmospheric oxygen, some nitrogen fixers adopted an anaerobic habitat and others developed heterocysts. By the time eukaryotes appeared, apparently more than half a billion years later, oxygen was abundant and fixed nitrogen (both NH_3 and NO_3^-) was probably less scarce, and so the eukaryotes never developed the enzymes needed for nitrogen fixation.

At present oxygen-releasing photosynthesis by green plants, cyanobacteria and some protists is responsible for the synthesis of most of the world's organic matter. It is not, however, the only mechanism of photosynthesis. The alternative systems are confined to a few groups of bacteria that on a global scale seem to be of minor importance today but that may have been far more significant in the geological past.

The several groups of photosynthetic bacteria differ from one another in their pigmentation, but they are alike in one important respect: unlike the photosynthesis of cyanobacteria and eukaryotes, all bacterial photosynthesis is a totally anaerobic process. Oxygen is not given off as a by-product of the reaction, and the photosynthesis cannot proceed in the presence of oxygen. Whereas oxygen appears to be required in green plants for the synthesis of chlorophyll *a*, oxygen inhibits the synthesis of bacteriochlorophylls.

The anaerobic nature of bacterial photosynthesis seems to present a paradox: photosynthetic organisms thrive where light is abundant, but such environments are also generally ones having high concentrations of oxygen, which poisons bacterial photosynthesis. These contradictory needs can be explained if it is assumed that anaerobic photosynthesis evolved among primitive bacteria early in the Precambrian, when the atmosphere was essentially anoxic. The photosynthesizers could thus have lived in matlike communities in shallow water and in full sunlight. Somewhat later such bacteria gave rise to the first organisms capable of aerobic photosynthesis, the precursors of modern cyanobacteria. For the anaerobic photosynthetic bacteria the moleculary oxygen released by this mutant strain was a toxin, and as a result the aerobic photosynthesizers were able to supplant the anaerobic ones in the upper portions of the mat communities. The anaerobic species became adapted to the lower parts of the mat, where there is less light but also a lower concentration of oxygen. Many photosynthetic bacteria occupy such habitats today.

Photosynthetic bacteria were surely not the first living organisms, but the history of life in the period that preceded their appearance is still obscure. What little information can be inferred about that early period, however, is consistent with the idea that the environment was then largely anoxic. One tentative line of evidence rests on the assumption that among organisms living today those that are simplest in structure and in biochemistry are probably the most closely related to the earliest forms of life. Those simplest organisms are bacteria of the clostridial and methanogenic types, and they are all obligate anaerobes.

There is even a basis for arguing that anoxic conditions must have prevailed during the time when life first emerged on the earth. The argument is based on the many laboratory experiments that have demonstrated the synthesis of organic compounds under conditions simulating those of the primitive planet. These syntheses are inhibited by even small concentrations of molecular oxygen. Hence it appears that life probably would not have developed at all if the early atmosphere had been oxygen-rich. It is also significant that the starting materials for such experiments often include hydrogen sulfide and carbon monoxide (CO), and that an intermediate in many of the reactions is hydrogen cyanide (HCN). All three compounds are poisonous gases, and it seems paradoxical that they should be forerunners of the earliest biochemistry. They are poisonous, however, only for aerobic forms of life: indeed, for many anaerobes hydrogen sulfide not only is harmless but also is an important metabolite.

It was argued above that oxygen must have been freely available by the time the first eukaryotic cells appeared, probably 1,400 to 1,500 million years ago. Hence the proliferation of cyanobacteria that released the oxygen must have taken place earlier in the Precambrian. How much earlier remains in question. The best available evidence bearing on the issue comes from the study of sedimentary minerals, some of which may have been influenced by the concentration of free-oxygen at the time they were deposited. In recent years a number of workers have investigated this possibility, most notably Preston E. Cloud, Jr., of the University of California at Santa Barbara and the U.S. Geological Survey.

One mineral of significance in this argument is uraninite (UO_2), which is found in several deposits that were laid down in Precambrian streambeds. In the presence of oxygen, grains of uraninite are readily oxidized (to U_3O_8) and are thereby dissolved. David E. Grandstaff of Temple University has shown that streambed deposits of the mineral probably could not have accumulated if the concentration of atmospheric oxygen was greater than about 1 per cent. Uraninite-bearing deposits of this type are found in sediments older than about two billion years but not in younger strata, suggesting that the transition in oxygen concentration may have come at about that time.

Another kind of mineral deposit, the iron-rich formations called red beds, exhibits the opposite temporal pattern: red beds are known in sedimentary sequences younger than about two billion years but not in older ones. The red beds are composed of particles coated with iron oxides (mostly the mineral hematite, Fe_2O_3), and many are thought to have formed by exposure to oxygen in the atmosphere rather than under water. It has been proposed that the oxygen may have been biologically generated. This hypothesis is consistent with several lines of evidence, but objections to it have also been raised. For example, most red beds are continental deposits rather than marine ones and are therefore susceptible to erosion; it is thus conceivable that red beds were formed earlier than two billion years ago as well as later but that the earlier beds have been destroyed. It is also possible that the oxygen in the red beds had a nonbiological origin; it may have come from the splitting of water by ultraviolet radiation. This has apparently happened on Mars to create a vast red bed across the surface of that planet, where there are only traces of free oxygen and there is no evidence of life.

Perhaps the most intriguing mineral evidence for the date of the oxygen transition comes from another kind of iron-rich deposit: the banded iron formation. These deposits include some tens of billions of tons of iron in the form of oxides embedded in a silica-rich matrix; they are the world's chief economic reserves of iron. A major fraction of them was deposited within a comparatively brief period of a few

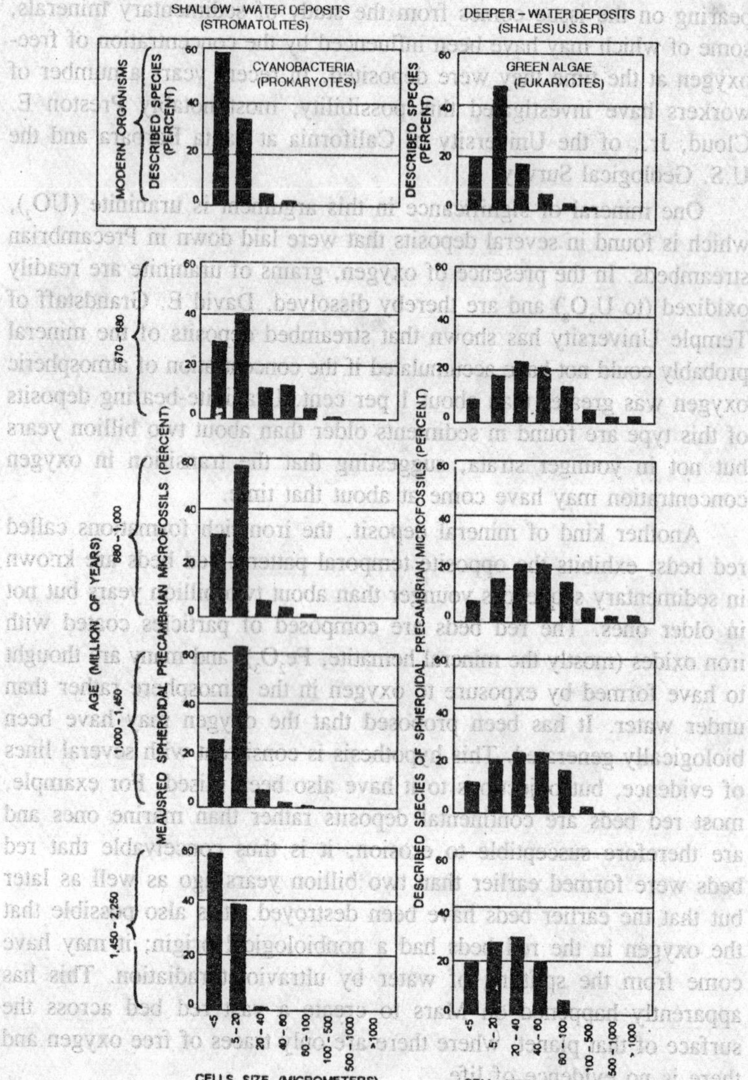

Fig. 4.5. Size of fossil cells provides evidence on the origin of the eukaryotes.

hundred million years beginning somewhat earlier than two billion years ago.

A transition in oxygen concentration could explain this major episode of iron sedimentation through the following hypothetical sequence of events. In a primitive, anoxic ocean, iron existed in the

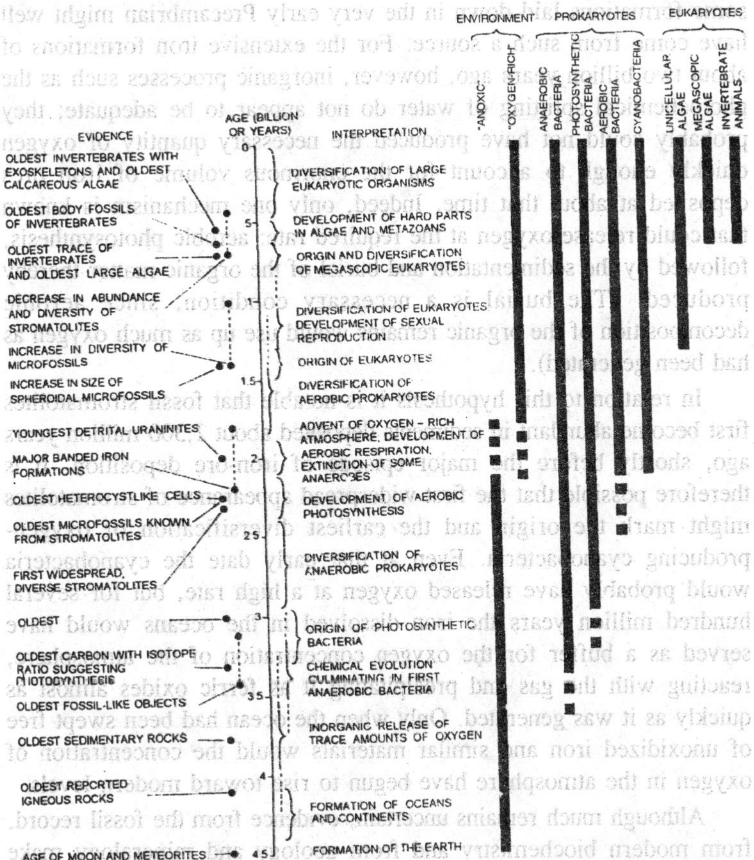

Fig. 4.6. Major events in Precambrian evolution are presented in chronological sequence based on evidence from the fossil record, from inorganic geology and from comparative studies of the metabolism and biochemistry of modern organisms.

ferrous state (that is, with a valence of +2) and in that form was soluble in seawater. With the development of aerobic photosynthesis small concentrations of oxygen began diffusing into the upper portions of the ocean, where it reacted with the dissolved iron. The iron was thereby converted to the ferric form (with a valence of +3), and as a result hydrous ferric oxides were precipitated and accumulated with silica to form rusty layers on the ocean floor. As the process continued virtually all the dissolved iron in the ocean basins was precipitated: in a matter of a few hundred million years the world's oceans rusted.

As in the deposition of red beds, an inorganic origin could also be proposed for the oxygen in the banded iron formations; the oxygen in some formations laid down in the very early Precambrian might well have come from such a source. For the extensive iron formations of about two billion years ago, however, inorganic processes such as the photochemical splitting of water do not appear to be adequate; they probably could not have produced the necessary quantity of oxygen quickly enough to account for the enormous volume of iron ores deposited at about that time. Indeed, only one mechanism is known that could release oxygen at the required rate: aerobic photosynthesis, followed by the sedimentation and burial of the organic matter thereby produced. (The burial is a necessary condition, since aerobic decomposition of the organic remains would use up as much oxygen as had been generated).

In relation to this hypothesis it is notable that fossil stromatolites first become abundant in sediments deposited about 2,300 million years ago, shortly before the major episode of iron-ore deposition. It is therefore possible that the first widespread appearance of stromatolites might mark the origin and the earliest diversification of oxygen-producing cyanobacteria. Even at that early date the cyanobacteria would probably have released oxygen at a high rate, but for several hundred million years the iron dissolved in the oceans would have served as a buffer for the oxygen concentration of the atmosphere, reacting with the gas and precipitating it as ferric oxides almost as quickly as it was generated. Only when the ocean had been swept free of unoxidized iron and similar materials would the concentration of oxygen in the atmosphere have begun to rise toward modern levels.

Although much remains uncertain, evidence from the fossil record, from modern biochemistry and from geology and mineralogy make possible a tentative outline for the history of precambrian life. The most primitive forms of life with recognizable affinities to modern organisms were presumably spheroidal prokaryotes, perhaps comparable to modern bacteria of the clostridial type. Initially at least they probably derived their energy from the fermentation of materials that were organic in nature but were of nonbiological origin. These materials were synthesized in the anoxic early atmosphere and were of the type that during the age of chemical evolution had led to the development of the first cells.

The first photosynthetic organisms apparently arose earlier than about three billion years ago. They were anaerobic prokaryotes, the precursors of modern photosynthetic bacteria. Most of them probably lived in matlike communities in shallow water, and they may have been responsible for building the earliest fossil stromatolites known, which are estimated to be about three billion years old.

The rise of aerobic photosynthesis in the mid-Precambrian introduced a change in the global environment that was to influence all subsequent evolution. The resulting increase in oxygen concentration probably led to the extinction of many anaerobic organisms, and others were forced to adopt marginal habitats, such as the lower reaches of bacterial mat communities. Nitrogen-fixing organisms also retreated to anaerobic habitats or developed heterocyst cells. With little competition for those regions having optimum light the cyanobacteria were able to spread rapidly and came to dominate virtually all accessible habitats. With the development of the citric acid cycle and its more efficient extraction of energy from foodstuff, the dominance of the biological community by aerobic organisms was confirmed. When the major episode of deposition of banded iron formations ended some 1,800 million years ago, the trend toward increasing oxygen concentration became irreversible.

By the time eukaryotic cells arose 1,500 to 1,400 million years ago a stable, oxygen-rich atmosphere had long prevailed. Adaptive strategies needed by earlier organisms to cope with fluctuations in the oxygen level were unnecessary for eukaryotes, which were from the start fully aerobic. The diversity of eukaryote cell types present by about a billion years ago suggests that some form of sexual reproduction may have evolved by then. Within the next 400 million years the rapid diversification of eukaryotic organisms had led to the emergence of multicellular forms of life, some of them recognizable antecedents of modern plants and animals.

In style and in tempo evolution in the Precambrian was distinctly different from that in the later, Phanerozoic era. The Precambrian was an age in which the dominant organisms were microscopic and prokaryotic, and until near the end of the era the rate of evolutionary change was limited by the absence of advanced sexual reproduction. It was an age in which the major benchmarks in the history of life were the result of biochemical and metabolic innovations rather than of morphological changes. Above all, in the Precambrian the influence of life on the environment was at least as important as the influence of the environment on life. Indeed, the metabolism of all the plants and animals that subsequently evolved was made possible by the photosynthetic activities of primitive cyanobacteria some two billion years ago.

5

MICROMOLECULES

Water spews from the Earth as geysers. It circles the globe as clouds and falls from the sky as the gentle rain or in thundering torrents. Frozen, it covers parts of Antarctica to a depth of 3,000 meters (10,000 feet) or more. Vast oceans of it submerge much of the planet. Water is one of the key ingredients that makes life on Earth possible. It makes up as much as 95 percent of the weight of some living things.

Water is a simple substance containing only three atoms-two of hydrogen and one of oxygen. What does the composition of water tell us about its characteristics, and why is water so important to living systems? We can't answer these questions without more information about atoms in general, and about hydrogen and oxygen atoms in particular. But it is not only the constituents of matter that are important. To understand the Behaviour of something as apparently simple as water, we need to know how its constituent atoms are linked together. The same is true of the other small molecules that are essential to living systems.

The first part of this chapter will address the constituents of matter: atoms—their variety, properties, and capacity to combine with other atoms. Then we'll consider how matter changes. In addition to changes in state (solid to liquid to gas), substances undergo changes that transform both their composition and their characteristic properties. When cells use. oxygen to "burn" glucose, the products are water, carbon dioxide, and energy to power life activities. This transformation is similar to what happens when the fuel propane is burned in a stove, a combustion reaction. By studying simple systems such as the combustion of propane, we can understand better what happens in systems as complicated as living cells. The discussion of general chemical principles and their application to small molecules aids our understanding of the large molecules that form the basis for the life of an organism.

Later in this chapter, we return to a consideration of the structure and properties of water and its relationship to acids and bases. We close with a bridge to the next chapter—a consideration of characteristic groups of atoms that contribute specific properties to larger molecules of which they are part.

ATOMS: CONSTITUENTS OF MATTER

More than a million million (1×10^{12}) atoms could fit in a single layer over the period at the end of this sentence. Each atom consists of a dense, positively charged nucleus, around which one or more negatively charged electrons move. The nucleus contains one or more protons and may contain one or more neutrons. Atoms and their component particles have mass. Mass is a property of all matter. Measuring mass measures the quantity of matter present. The greater the mass, the greater the quantity of matter.

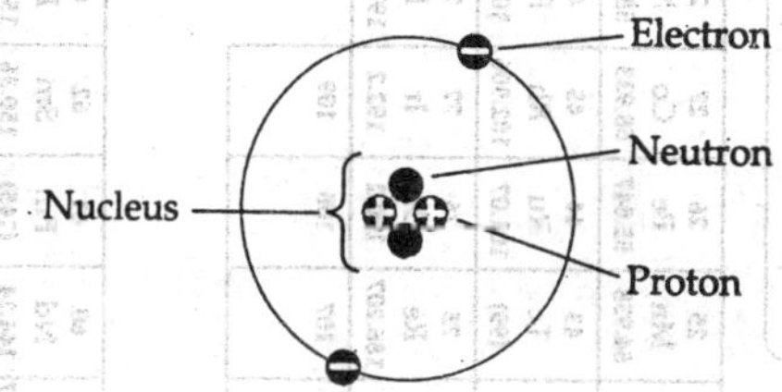

The mass of a proton serves as a standard unit: the atomic mass unit (amu), or dalton (named after the English chemist John Dalton). A single proton or neutron has the mass of 1 dalton, which is 1.7×10^{-24} grams (0.0000000000000000000000017 g). The mass of an electron is 9×10^{-28} g (0.0005 dalton). Because the mass of an electron is so much less than the mass of a proton or a neutron, the contribution of electrons to the mass of an atom can usually be ignored.

The positive electric charge on a proton is defined as a unit of charge. An electron has a charge equal and opposite to that of a proton. Thus the charge of a proton is +1 unit, that of an electron is -1 unit. Unlike charges attract each other; like charges repel. The neutron, as its name suggests, is electrically neutral, so its charge is 0 unit. Because the number of protons in an atom equals the number of electrons, the atom itself is electrically neutral.

An Element is Made Up of Only One Kind of Atom

The element hydrogen consists only of hydrogen atoms; the element iron consists only of iron atoms. An element is a pure substance that

The six elements highlighted in yellow make up 98% of the mass of any living organism.

Elements shown in orange are present in tiny amounts in many organisms.

Vertical columns have elements with similar properties.

Key: Chemical symbol / Atomic number / Atomic mass (e.g. 12 Mg 24.305)

1 H 1.0079																	2 He 4.003
3 Li 6.941	4 Be 9.012											5 B 10.81	6 C 12.011	7 N 14.007	8 O 15.999	9 F 18.998	10 Ne 20.179
11 Na 22.990	12 Mg 24.305											13 Al 26.982	14 Si 28.086	15 P 30.974	16 S 32.06	17 Cl 35.453	18 Ar 39.948
19 K 39.098	20 Ca 40.08	21 Sc 44.956	22 Ti 47.88	23 V 50.942	24 Cr 51.996	25 Mn 54.938	26 Fe 55.847	27 Co 58.933	28 Ni 58.69	29 Cu 63.546	30 Zn 65.38	31 Ga 69.72	32 Ge 72.59	33 As 74.922	34 Se 78.96	35 Br 79.909	36 Kr 83.80
37 Rb 85.4778	38 Sr 87.62	39 Y 88.906	40 Zr 91.22	41 Nb 92.906	42 Mo 95.94	43 Tc (99)	44 Ru 101.07	45 Rh 102.906	46 Pd 106.4	47 Ag 107.870	48 Cd 112.41	49 In 114.82	50 Sn 118.69	51 Sb 121.75	52 Te 127.60	53 I 126.904	54 Xe 131.30
55 Cs 132.905	56 Ba 137.34	57–71 La–Lu	72 Hf 178.49	73 Ta 180.948	74 W 183.85	75 Re 186.207	76 Os 190.2	77 Ir 192.2	78 Pt 195.08	79 Au 196.967	80 Hg 200.59	81 Tl 204.37	82 Pb 207.19	83 Bi 208.980	84 Po (209)	85 At (210)	86 Rn (222)
87 Fr (223)	88 Ra 226.025	89–103 Ac–Lr	104	105	106	107	108	109									

Lanthanide series	57 La 138.906	58 Ce 140.12	59 Pr 140.9077	60 Nd 144.24	61 Pm (145)	62 Sm 150.36	63 Eu 151.96	64 Gd 157.25	65 Tb 158.924	66 Dy 162.50	67 Ho 164.930	68 Er 167.26	69 Tm 168.934	70 Yb 173.04	71 Lu 174.97
Actinide series	89 Ac 227.028	90 Th 232.038	91 Pa 231.0359	92 U 238.02	93 Np 237.0482	94 Pu (244)	95 Am (243)	96 Cm (247)	97 Bk (247)	98 Cf (251)	99 Es (252)	100 Fm (257)	101 Md (258)	102 No (259)	103 Lr (260)

Fig. 5.1. The periodic table.

contains only one type of atom. The atoms of each element have certain characteristics or properties that distinguish them from the atoms of other elements. The more than 100 elements found in the universe are arranged in the periodic table. The periodic table arranges elements with similar properties in vertical columns in order of their increasing size. Although there are more than 100 elements in the world, about 98 percent of the mass of a living organism (bacterium, turnip, or human) is composed of just six elements—carbon, hydrogen, nitrogen, oxygen, phosphorus, and sulfur. The chemistry of these elements will be our primary concern.

A substance (such as oxygen gas) that contains only one kind of atom is an elemental substance. A substance that contains more than one kind of atom is a *compound.* Most substances of biological interest are compounds.

Number of Protons Identifies the Element

An atom is distinguished from other atoms by the number of its protons, which does not change. This number is called the atomic number. An atom of hydrogen contains 1 proton, a helium atom has 2 protons, carbon has 6 protons, and plutonium has 94. The atomic numbers of these elements are thus 1, 2, 6, and 94, respectively.

Every atom except hydrogen has one or more neutrons in its nucleus. The *mass number* of an atom equals the total number of protons and neutrons in its nucleus. Because the mass of an electron is infinitesimal compared with that of a neutron or proton, electrons are ignored in calculating the mass number. The nucleus of a helium atom contains 2 protons and 2 neutrons; oxygen has 8 protons and 8 neutrons. Helium, therefore, has a mass number of 4 and oxygen a mass number of 16. The mass number may be thought of as the weight of the atom, in daltons.

Each element has its own one- or two-letter symbol. For example, H stands for hydrogen, He for helium, and O for oxygen. Some symbols come from other languages: Fe (from Latin *ferrum*) stands for iron, Na (Latin *natrium*) for sodium, and W (German *Wolfram*) for tungsten. The periodic table gives the symbols for all of the 92 natural elements, as well as those for 14 elements that do not occur naturally.

Isotopes Differ in Number of Neutrons

We have been speaking of hydrogen and oxygen as if each had only one atomic form. But this is not true. Not all atoms of an element have the same mass number. The different atomic forms of a single element are called *isotopes* of the element. Isotopes of an element differ in the number of neutrons in the atomic nucleus. The common form of hydrogen is ^{1}H, but about one out of every 6,500 hydrogen atoms on Earth has a neutron as well as a proton in its nucleus and is thus ^{2}H, called deuterium. Furthermore, it is possible to create ^{3}H, tritium, which has *two* neutrons and a proton in its nucleus. Because all three types of hydrogen atoms have only one proton, they all have the atomic number 1. Deuterium, tritium, and common hydrogen have virtually identical chemical properties, although ^{2}H is twice and ^{3}H three times as heavy as ^{1}H.

In nature, many elements exist as several isotopes. For example, the natural isotopes of carbon are ^{12}C, ^{13}C, and ^{14}C. Unlike the hydrogen isotopes, the isotopes of most other elements do not have distinct names. Rather they are written in the form shown here and are referred to as carbon-12, carbon-13, and carbon-14, respectively. Most carbon atoms are ^{12}C, about 1.1 percent are ^{13}C, and a tiny fraction are ^{14}C. An element's atomic *mass* (*atomic* weight) is the average of the mass numbers of a representative sample of atoms of the element, with all isotopes in their normal proportions. For example, the atomic mass of carbon is 12.011. In biology, one encounters the terms "weight" and "atomic weight" more frequently than "mass" and "atomic mass"; therefore, we will use "weight" for the remainder of this book. Thus we say that the atomic weight of carbon is 12.011.

Some isotopes, called *radioisotopes*, are unstable and spontaneously give off energy as α (alpha), β (beta), or γ (gamma) radiation from the atomic nucleus. Such radioactive decay transforms the original atom into another type, usually of another element. For example, uranium-238 loses an alpha particle to form thorium-234, and carbon-14 loses a beta particle to form nitrogen-14. Biologists can incorporate radioisotopes into molecules and use the emitted radiation as a tag to identify changes that the molecules undergo in the body or to identify the locations of molecules within the cell. Some radioisotopes commonly used in biological experiments are ^{3}H (tritium), ^{14}C (carbon-14), and ^{32}P (phosphorus-32).

Although radioisotopes are useful for experiments and medicine, even low doses of radiation from radioisotopes have the potential to damage molecules and cells. Gamma radiation from cobalt-60 (^{60}Co) is used medically to damage or kill rapidly dividing cancer cells. In addition to these applications, radioisotopes can be used to date fossils.

Electron Behaviour Determines Chemical Bonding

In atoms, biologists are concerned primarily with electrons. To understand organisms, biologists study chemical changes that occur in living cells. These changes, called *chemical reactions* or just *reactions*, are changes in the atomic composition of substances. They occur because of the way in which electrons behave. The characteristic number of electrons in each atom of an element determines how the atom reacts with other atoms. All chemical reactions involve changes in the relationships of electrons with each other.

The location of a given electron in an atom at any given time is impossible to determine. We can only describe a volume of space within the atom where the electron is likely to be. The region of space within which the electron is found at least 90 percent of the time is the electron's orbital. In an atom, a given orbital can be occupied by at most two electrons. Thus any atom larger than helium (atomic number 2) must have electrons in two or more orbitals. The different orbitals have characteristic forms and orientations in space.

The orbitals constitute a series of *electron shells*, or energy levels, around the nucleus. The innermost electron shell, called the *K* shell, consists of only one orbital, called an *s* orbital. The *s* orbital fills first, and its electrons have the lowest energy. Hydrogen ($_1H$) has one *K* shell electron; helium ($_2He$) has two. All other atoms have two *K* shell electrons, as well as electrons in other shells. The *L* shell is made up of four orbitals (an *s* orbital and three *p* orbitals) and hence can hold up to eight electrons. The M, N, O, P, and Q shells have different numbers of orbitals, but the outermost orbitals can usually hold only eight electrons.

In any atom, the outermost shell determines how the atom combines with other atoms; that is, these outermost electrons determine how an atom behaves chemically. When an outermost shell consisting of four orbitals contains eight electrons, the atom is stable and will

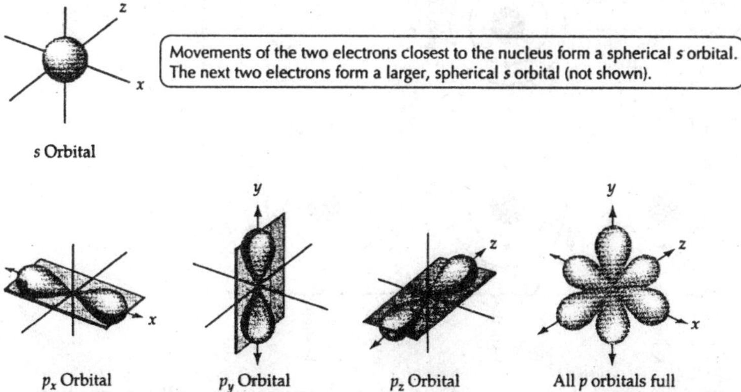

Fig. 5.2. Electron orbitals.

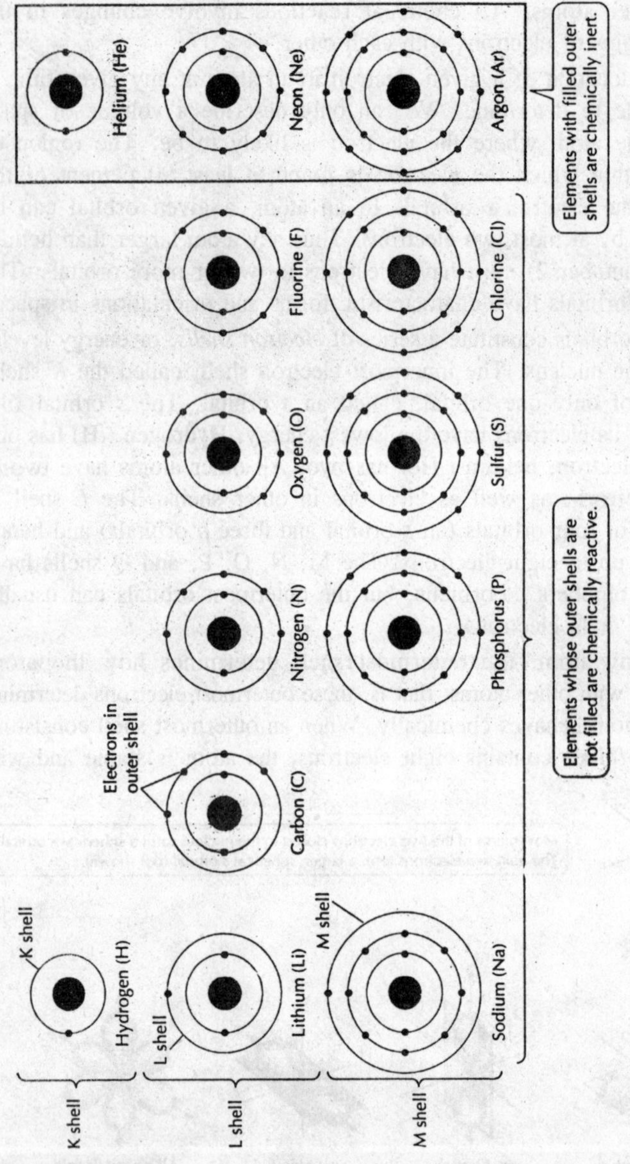

Fig. 5.3. Electron shells determine the reactivity of atoms.

not react with other atoms. Examples of some chemically inert elements are neon and argon, in each of which the outermost shell contains eight electrons. The atoms of other elements seek to attain the stable condition of having filled outer orbitals. They attain this stability by sharing electrons with other atoms or by gaining or losing one or more electrons from their outermost orbitals.

When they share electrons, atoms are bonded together. Such bonds create stable associations of atoms called molecules. A *molecule* can be defined as two or more atoms linked by chemical bonds. The tendency of atoms in stable molecules to have eight electrons in their outermost orbitals is known as the *octet rule*.

Many atoms in biologically important molecules follow the octet rule-for example, carbon (C) and nitrogen (N). However, some biologically important atoms such as hydrogen and phosphorus are exceptions to the rule. Hydrogen (H) attains stability when two electrons occupy its outermost orbital; phosphorus (P) is stable when its outermost orbitals contain ten electrons.

Chemical Bonds: Linking Atoms Together

A *chemical bond* is an attractive force that links two atoms to form a molecule. There are different kinds of chemical bonds, but all strong chemical bonds result from an atom's tendency to attain stability by filling its outermost electron orbitals. Atoms can gain stability in the outermost orbitals by sharing electrons or by losing or gaining one or more electrons. In this section, we will first discuss covalent bonds, the strong bonds that result from sharing of electrons. Then we'll examine hydrogen bonds, which are weaker than covalent bonds but enormously important to biology. Finally, we'll consider ionic bonding, which results when ions form as a consequence of the complete loss or gain of electrons by atoms.

Covalent Bonds Consist of Shared Pairs of Electrons

When two atoms attain stable electron numbers in their outer shells by sharing one or more pairs of electrons, a *covalent bond* forms. A hydrogen atom has one electron in its only shell, but two electrons would be a more stable condition. Imagine two hydrogen atoms, initially far apart but coming closer and closer, until they begin to interact. The negatively charged electron in each hydrogen atom is attracted by the positively charged proton in the nucleus of the other hydrogen atom. When the two atoms are close enough, the two electrons spend time between both nuclei, and the two atoms are covalently bonded together, forming a molecule of hydrogen gas (H_2). The two hydrogen nuclei share the two electrons equally and completely.

The two atoms do not come *too* close together, because their positively charged nuclei strongly repel each other. A certain distance between the coupled atoms gives the most stable arrangement. Pulling the atoms slightly farther apart would require an input of energy because of the "gluing" effect of the shared electrons. Pushing the atoms closer together would require energy because of the mutual repulsion of the protons. So the most stable arrangement of the covalently bonded hydrogens can also be described as an arrangement that has a minimum amount of energy and that is less reactive than are the individual atoms alone, each of which has an incompletely filled orbital in the K shell.

A carbon atom has a total of six electrons; two electrons fill its inner shell and four are in its outer L shell. Because the L shell can hold up to eight electrons, this atom can share electrons with up to four other atoms. Thus it can form four covalent bonds. When an atom of carbon reacts with four hydrogen atoms, a substance called methane (CH_4) forms, resulting from the overlapping of electron orbitals. Thanks to electron sharing, the outer shell of methane's carbon atom is filled with eight electrons, and the outer shell of each hydrogen atom is also filled. Thus four covalent bonds—each bond consisting of a shared pair of electrons—hold methane together.

Orientation of Bonds in Space

Not only the number, but also the spatial orientation, of bonds is important. The four filled orbitals around the carbon nucleus of methane distribute themselves in space so that the bonded hydrogens are directed to the corners of a regular tetrahedron with carbon in the center. Although the orientation of orbitals and shapes of molecules differ depending on the kinds of atoms and how they are linked together, it is essential to remember that all molecules occupy space and have three-dimensional shapes. The shapes of molecules contribute to their biological functions.

Multiple Covalent Bonds

A covalent bond is represented by a line between the chemical symbols for the atoms. Bonds in which a single pair of electrons is shared are called *single bonds* (for example, H-H, C-H). When four electrons (two pairs) are shared, the link is a *double bond* (C=C). In the gas ethylene ($H_2C=CH_2$), two carbon atoms share two pairs of electrons. *Triple bonds* (six shared electrons) are rare, but there is one in nitrogen gas (N-N), the chief component of the air we breathe. In the covalent bonds in these five examples, the electrons are shared more or less equally between the nuclei; consequently all regions of

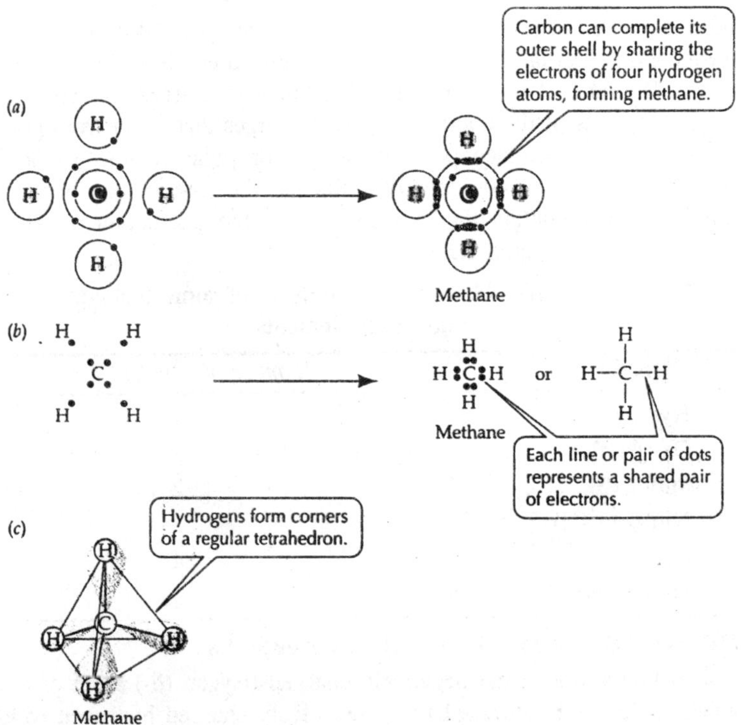

Fig. 5.4. Electrons are shared in covalent bonds.

the bonds are identical. However, when electrons are shared unequally in a covalent bond, regions of partial electric charge exist.

Uniqual Sharing of Electrons

So far we have discussed the covalent bonds that result from the equal sharing of electrons between two nuclei. Now we want to consider the kind of covalent bond that results from unequal sharing of the electrons.

Some atoms hold electrons to themselves more firmly than other atoms do. This characteristic is called *electronegativity*. Highly electronegative atoms that form covalent bonds include oxygen and nitrogen. When these atoms are covalently bonded to atoms with weaker electronegativity, such as carbon and hydrogen, the bonding pair of electrons is unequally shared between the two atoms, and the result is a *polar covalent bond*. For example, when oxygen is bonded to hydrogen, the bonding electrons spend much more time near the oxygen nucleus than near the hydrogen nucleus. Consequently, the oxygen end of the bond is slightly negative (symbolized δ- and spoken as "delta negative," meaning a partial unit charge), and the hydrogen end is slightly positive (S+). The bond is polar because these opposite charges are separated at the two ends of the bond. The partial charges that result from polar covalent bonds produce polar *molecules* or polar regions of large molecules. Polar bonds greatly influence the interactions between molecules that contain them, as we see in the interaction of water molecules in the liquid state.

Table 5.1. Typical bonding capabilities of some biologically important elements

Elements	*Number of covalent bonds*
Hydrogen (H)	1
Oxygen (O)	2
Sulfur (S)	2
Nitrogen (N)	3
Carbon (C)	4
Phosphorus (P)	5

Hydrogen Bonds may Form between Molecules

In liquid water, the negatively charged oxygen (δ-) atom of one water molecule is attracted to the positively charged hydrogen (δ+) of another water molecule. (Remember, negative charges attract positive charges.) The bond resulting from this attraction is called a *hydrogen bond* and is usually symbolized by a series of dots. Hydrogen bonds are not restricted to water molecules. They may form between any covalently bonded hydrogen and an electronegative atom, usually oxygen or nitrogen: -H···O- or -H ···N-. Hydrogen bonds form between small molecules or between different parts of large molecules. Covalent bonds and polar covalent bonds, on the other hand, are always found *within* molecules.

A hydrogen bond is a weak bond; it has about onetwentieth (5 percent) of the strength of a covalent bond between a hydrogen atom and an oxygen atom. However, where many hydrogen bonds form, they have considerable strength and greatly influence the structure and properties of substances. Later in this chapter we'll discuss further how hydrogen bonding in water contributes to many of the properties of water that are significant for living systems. Hydrogen bonds also play important roles in determining and maintaining the three-dimensional shapes of giant molecules such as DNA and protein.

Ions Form Bonds by Electrical Attraction

When one interacting atom is much more electronegative than the other, a complete transfer of one or more electrons may take place. For example, a sodium atom has only one electron in its outermost shell; this condition is unstable. A chlorine atom has seven electrons in its outer shell, another unstable condition. The reaction between sodium and chlorine makes both atoms more stable. When the two atoms meet, the highly *electronegative chlorine atom takes the single unstable* electron from the sodium atom. The result is two electrically charged particles, called *ions*. Ions are electrically charged particles that form when atoms gain or lose one or more electrons.

The sodium ion (Na^+) has a +1 unit charge because it has one less electron than it has protons. The outermost electron shell of the sodium ion is full, with eight electrons, so the ion is stable. The chloride ion (Cl^-) has a -1 unit charge because it has one more electron than it has protons. This additional electron gives Clan outer shell with a stable load of eight electrons. Negatively charged ions are called *anions*; positively charged ions are called *cations*.

Some elements form ions with multiple charges by losing or gaining more than one electron to achieve a stable electron configuration in their outer shell. Examples are Ca^{2+} (the calcium ion, created from a calcium atom that has lost two electrons), Mg^+ (magnesium ion), and Al^{3+} (aluminum ion). Two biologically important elements each yield more than one stable ion: Iron yields Fee^+ (ferrous ion) and Fe^{3+} (ferric ion), and copper yields Cu^+ (cuprous ion) and Cue^+ (cupric ion). Groups of covalently bonded atoms that carry an electric charge are called *complex ions*; examples include NH_4^+ (ammonium ion), SO (sulfate ion), and PO_4^{2-} (phosphate ion).

Once they form, ions 'are usually stable, and no more electrons are lost or gained. As stable entities, ions can enter into stable associations through ionic bonding. Thus stable solids such as sodium chloride (NaCl) and potassium phosphate (K_3P0_4) are formed: Although a very complex solid, bone has as one of its major components the simple ionic compound $Ca_3(PO_4)_2$.

Ionic Bonds: Electrical Attraction

Ionic bonds are the bonds formed by electrical attractions between ions bearing opposite charges. In solids such as table salt (NaCl), the cations and anions are held together by ionic bonds. In solids, the ionic bonds are strong because the ions are close together. However, when ions are dispersed in water, the distance between them can be

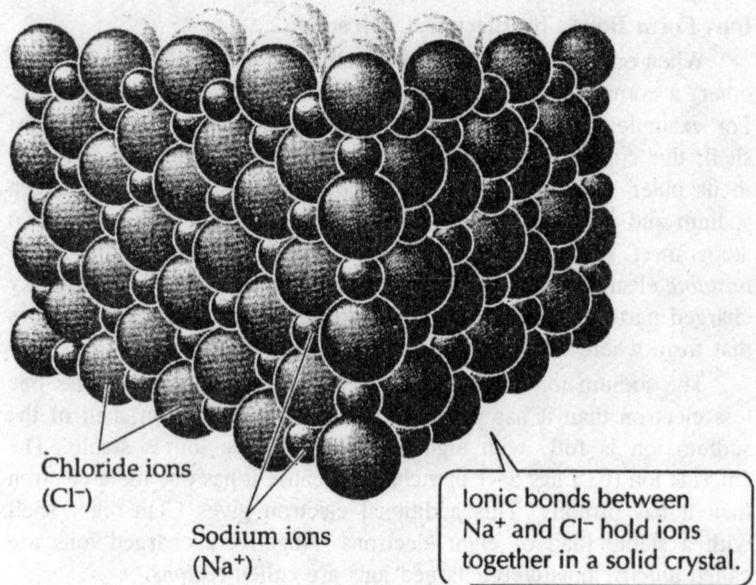

Fig. 5.5. Ionic binding in a solid.

large; the strength of their attraction is thus greatly reduced. Under the conditions that exist in the cell, an ionic bond is less than one-tenth as strong as a covalent bond that shares electrons equally, so an ionic bond can be broken much more readily than a covalent bond.

Not surprisingly, ions with one or more unit charges can interact with polar substances as well as with other ions. Such interaction results when table salt or any other ionic solid dissolves in water. The hydrogen bond that we described earlier is a weak type of ionic bond, because it is formed by electrical attractions. However, it is weaker than most ionic bonds because the hydrogen bond is formed by partial charges (δ- and δ+) rather than by whole unit charges (+1 unit, –1 unit).

Nonpolar Substances have no Attraction for Polar Substances

We have been discussing the bonds that result from electrical attractions between positive and negative charges (ionic bonds and hydrogen bonds). Now let's return to a brief consideration of substances that have "pure" covalent bonds.

These bonds form between atoms that have equal or nearly equal electronegativities-such as carbon and hydrogenwhich share the bonding electrons equally. Such bonds are abundant in the compounds of hydrogen

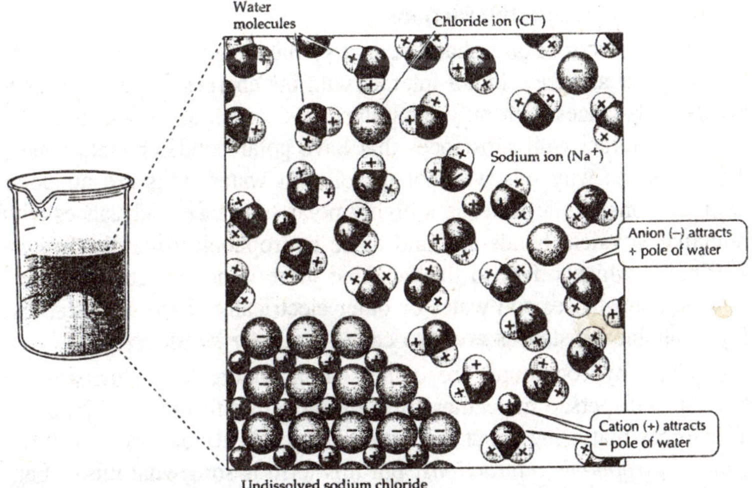

Fig. 5.6. Water molecules surrounded ions.

and carbon—the hydrocarbons. Molecules such as ethane (CH_3—CH_3) and butane (CH_3—CH_2—CH_2—CH_3) are small hydrocarbons, but in living systems, molecules exist with hydrocarbon chains consisting of 16 or more carbon atoms.

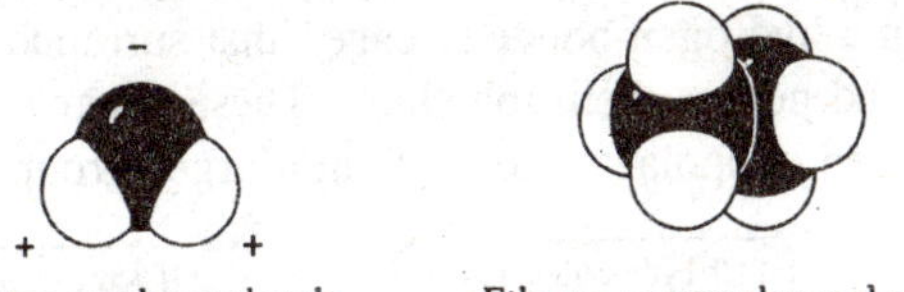

Fig. 5.7. Nonpolar molecules.

Attraction between Nonpolar Molecules

Nonpolar substances such as oils and fats show *van der Waals attractions* between molecules. These attractive forces operate only when nonpolar substances come very close to each other. The random variations in the electron distribution in one molecule create an opposite charge distribution in the adjacent molecule, and the result is a brief, weak attraction. Although each such interaction is brief and weak at any one site, the summation of many such interactions over the entire span of a nonpolar molecule can produce substantial attraction. Thus van der Waals interactions are important in holding together the long hydrocarbon chains that make up the inner portion of biological membranes. They also stabilize portions of the DNA double helix and the intricate folded structure of proteins.

Polar and Nonpolar Interactions

When electrons are shared equally, the resultant covalent bonds are nonpolar and they do not interact with the charges of polar covalent bonds. Substances with only nonpolar bonds, such as butane and oils, will not interact with substances that have polar bonds or ionic bonds. This explains why oils will not dissolve in water. Oils are nonpolar hydrocarbons, while water is a highly polar substance. Substances with nonpolar covalent bonds are said to be hydrophobic (literally "water-fearing"), which refers to the fact that there is no attraction between nonpolar substances and water or other electrically charged substances. Hydrophobic substances are also called *nonpolar substances*.

When hydrocarbons are dispersed in water, they slowly come together—dispersed molecules form droplets that form larger droplets. The forces that bring about this combining of molecules are sometimes called *hydrophobic interactions*, but this term is somewhat misleading. The interactions that combine nonpolar substances have less to do with forces between the nonpolar molecules than with the hydrogen bonding of the water that surrounds the molecules.

When nonpolar substances such as hydrocarbons are introduced into water, they cause a disruption in the usual hydrogen bonding between water molecules. In the vicinity of the hydrocarbon, the water molecules form a hydrogen-bonded "cage" that surrounds the nonpolar hydrocarbons and pushes them together. These water cages can bring together dispersed nonpolar molecules into larger groups.

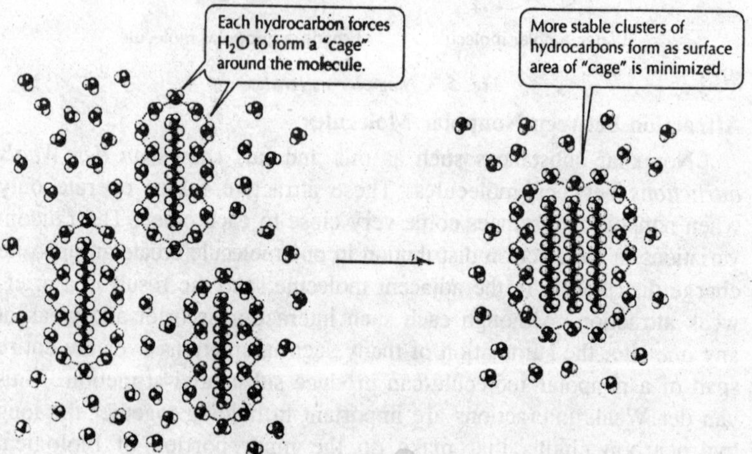

Fig. 5.8. Interaction of water and nonpolar substances.

Eggs by the Dozen, Molecules by the Mole

In modern biology, as in chemistry, the question "How much?" is as important as "What kind?" In this section, we will examine briefly how chemists and biologists deal quantitatively with atoms and molecules.

The molecular formula uses chemical symbols to identify different atoms, and subscript numbers to show how many atoms are present. For example, the molecular formula for methane is CH_4 (each molecule contains one carbon atom and four hydrogen atoms), that for oxygen gas is O_2, and that for sucrose (table sugar) is $C_{12}H_{22}O_{11}$. The hormone insulin is represented by the molecular formula $C_{2}5_{4}H_{377}N_{65}O_{76}S_{6}$! Although molecular formulas tell us what kinds of atoms and how many of each kind are present in the molecule, they tell us nothing about which atoms are linked to which. *Structural formulas* give us this information.

Each compound has a *molecular weight* (molecular mass): the sum of the atomic weights of the atoms in the molecule. The atomic weights of hydrogen, carbon, and oxygen are 1.008, 12.011, and 16.000, respectively. Thus the molecular weight of water (H_2O) is $(2 \times 1.008) + 16.000 = 18.016$, or about 18. What is the molecular weight of sucrose ($C_{12}H_{22}O_{11}$)? Your calculations should tell you that the answer is approximately 342. If you remember the molecular weights of a few representative biological compounds, you will be able to picture the relative sizes of molecules that interact with one another. Experiments require quantitative information, Suppose we want to compare how sodium chloride (NaCJ), potassium chloride (KCI), and lithium chloride (LiCI) affect a biological process. At first you might think we could simply give, say, 2 grams (g) of NaCl to one set of subjects, 2 g of KCI to another, and 2 g of LiCI to the third. But because the molecular weights of NaCl, KCI, and LiCI are different, 2-gram samples of each of these substances contain different numbers of molecules. The comparison would thus not be legitimate. Instead, we want to give *equal numbers of molecules* of each substance so that we can compare the activity of one molecule of one substance with that of one molecule of another. How can we measure out equal numbers of molecules?

We can Calculate Numbers of Molecules by Weighing

Measuring the number of molecules is essential to studying how many molecules take part in chemical reactions. Consider two large barrels, one filled with bolts and the other with pennies. How can we determine the number of units in each barrel? We could count them, but that would be tedious (and it is impossible to count individual molecules directly). However, if we know the weight of one bolt and one penny, we can calculate the number of units by dividing the total weight by the weight of each unit. The same principles apply to determining the number of molecules in a weighed sample.

To determine the number of molecules in a sample of a pure substance, we first determine the weight of the substance in grams, then we divide the grams by the relative weight of one molecule (the molecular weight defined earlier). Therefore, to measure out quantities of substances containing equal numbers of molecules, we weigh out in grams a quantity of each substance equivalent to its molecular weight. The molecular weight of NaCl is 58.45. Therefore 58.45 g of NaCl will have the same number of molecules as 74.55 g of KCI, whose molecular weight is 74.55.

If we divide the 74.55 g by the molecular weight of 74.55, the answer is 1. But one what? Surely not one molecule! The calculated numbers here are not the exact number, because we did not divide by the actual weight in grams of a molecule of potassium chloride (which would be about 1×10^{-22} g). Instead we divided by the relative weight. Therefore, instead of knowing the exact number of molecules, we know the relative number, which is called a *mole. One mole of a substance is an amount whose weight in grams is numerically equal to the molecular weight of the substance.* Potassium chloride (KCI) has a molecular weight of 74.55, so 1 mole of KCI weighs 74.55 g. Likewise, 1 mole of NaCl weighs 58.45 g, and 1 mole of LiCl, 42.40 g.

A mole of one substance contains the same number of molecules as does a mole of any other substance. This number, known as *Avogadro's number*, is 6.023×10^{23} molecules per mole. The concept of the mole is important for biology because it enables us to work easily with known numbers of molecules. Since we can neither weigh nor count individual molecules, we work with moles. The mole concept is analogous to the concept of a dozen or a gross. We buy some things, such as eggs, by the dozen (or another similar unit, such as the gross) because the individual items are inconvenient to count. Just as a dozen of anything contains twelve of that thing, a mole of any substance contains Avogadro's number of molecules of that substance.

Reactions Take Place in Solutions

In cells and in the laboratory, reactions take place in solutions formed when substances (solutes) are dissolved in water (the solvent).

The amount of substance dissolved in a given amount of water is the concentration of the solution. Knowing the concentrations of solutions is essential in performing experiments and interpreting their results.

A solution that contains 1 mole of solute per liter of solution has a one-molar concentration, abbreviated 1 *M*. A solution containing a half mole per liter is referred to as 0.5 *M*, or half-molar. How would you make 100 milliliters (ml) of a 0.5 *M* sucrose solution? The molecular weight of sucrose is 342, so 1 liter (1,000 ml) of a *1 M* sucrose solution contains 1 mole, or 342 g, of sucrose. You were asked to make just 100 ml of 0.5 *M* solution. Since 34.2 g of sucrose would make 100 ml of *1 M* sucrose, to make 100 ml of a 0.5 *M* sucrose solution, you would use 0.5 × 34.2 g = 17.1 g of sucrose.

Chemical Reactions: Atoms Change Partners

When atoms combine or change bonding partners, a chemical reaction is occurring. Consider the combustion reaction that takes place in the flame of a propane stove. When propane (C_3H_8) reacts with oxygen gas (O_2), the carbon atoms become bonded to oxygen atoms instead of to hydrogen atoms, and the hydrogen atoms become bonded to oxygen instead of to carbon. As the covalently bonded atoms change bonding partners, the composition of the matter changes, and propane and oxygen gas become carbon dioxide and water. This chemical reaction can be represented by the balanced equation

$$C_3H_8 + 5O_2 \rightarrow 3CO_2 + 4H_2O$$

In this equation, the propane and oxygen are the *reactants*, and the carbon dioxide and water are the *products*. The arrow symbolizes the chemical reaction. The numbers preceding the molecular formulas balance the equation and indicate how many molecules react or are produced. In this and all other chemical reactions, matter is neither created or destroyed. The total number of carbons on the left equals the total number on the right. However, there is another product of this reaction: energy.

The heat of the stove's flame and its blue light reveal that the reaction of propane and oxygen releases a great deal of energy. *Energy* is defined as the capacity to do work, but on a more intuitive level, it can be thought of as the capacity to change. Chemical reactions do not create or destroy energy, but *changes* in energy usually accompany chemical reactions. The energy released as heat and light was present in the reactants in another form, called *potential chemical energy*. In some chemical reactions, energy must be supplied from the environment (for example, some substances will react only after being heated), and some of this supplied energy becomes stored as potential chemical energy in the bonds formed in the reactants.

We can measure the energy associated with chemical reactions using the unit called a calorie (cal). A calorie is the amount of heat energy needed to raise the temperature of 1 g of pure water from 14.5°C to 15.5°C. (The nutritionist's Calorie, with a capital C, is what biologists call a kilocalorie (kcal) and is equal to 1,000 heat-energy calories.) Another unit of energy that is increasingly used is the joule (J). When you compare data on energy, always compare joules to joules and calories to calories. The two units can be interconverted: 1 J = 0.239 cal, and 1 cal = 4.184 J. Thus, for example, 486 cal = 2033 J, or 2.033 kJ. Although defined in terms of heat, the calorie and the joule are measures of any form of energy-mechanical, electric, or chemical.

Within living cells, chemical reactions called oxidation-reduction reactions take place that have much in common with this combustion of propane. The fuel for these biological reactions is different (the sugar glucose, rather than propane), and the reactions proceed by many intermediate steps that permit the energy released from the glucose to be harvested and put to use by the cell. But the products are the same: carbon dioxide and water. We will present and discuss oxidation-reduction reactions and several other types of chemical reactions that are prevalent in living systems in the chapters that follow.

Water: Structure and Properties

Water, like all other matter, can exist in three states: solid (ice), liquid, and gas (vapour). Liquid water is the medium in which life originated on Earth more than 3.5 billion years ago, and it is in water that life evolved for about a billion years. Today water covers three-fourths of Earth's surface, and all active organisms contain between 45 and 95 percent water. No organism can remain biologically active without water, which is important in both the outer and inner environments. Within cells, water participates directly in many chemical reactions, and it is the medium (or solvent) in which most reactions take place. In this section we will consider the structure and interactions of water molecules, exploring how these generate properties essential to life.

Water has a Unique Structure and Special Properties

The shape of a water molecule is determined by the distribution in space of the four pairs of electrons in the outer shell of the oxygen atom. Each pair of electrons is confined to an orbital. Two of these pairs are bonded to hydrogens, but the other two pairs (nonbonding pairs) also influence shape. Because of their negatively charged electrons, the four orbitals, each containing a pair of electrons, repel one another. In seeking to be as far apart as possible, they give the water molecule a nearly tetrahedral shape.

The shape of the water molecule, its polar nature, and the formation of hydrogen bonds between water molecules or with other substances give water its unusual properties. For example, ice floats, and compared to other liquids, water is an excellent solvent, making it an ideal medium for biochemical reactions. Water is both cohesive (sticking to itself) and adhesive (sticking to other things). And the energy changes that

Ice Floats

In its solid state (ice), water is held by its hydrogen bonds in a rigid, crystalline structure in which each water molecule is hydrogen-bonded to four other molecules. Although these molecules are held firmly in place, they are not as tightly packed as they are in liquid water. In other words, *solid water is less dense than liquid water,* which is why ice floats in water. If ice sank in water, as almost all other solids do in their corresponding liquids, ponds and lakes would freeze from the bottom up, becoming solid blocks of ice in winter and killing most of the organisms living in them. Once the whole pond had frozen, its temperature could drop well below the freezing point of water. However, because ice floats, it forms a protective insulating layer on the top of the pond, reducing heat flow to the cold air above. Thus fish, plants, and other organisms in the pond can survive the winter at temperatures no lower than 0°C, the freezing point of pure water.

Melting and Freezing

Compared to other nonmetallic substances of the same size, ice requires a great deal of heat energy to melt. Melting 1 mole of water molecules requires the addition of 5.9 kJ of energy. This value is high because more than a mole of hydrogen bonds must be broken for 1 mole of water to change from solid to liquid. In the opposite process, freezing, a great deal of energy must be lost for water to transform from liquid to solid. These properties help make water a moderator of temperature changes.

Another property of water that moderates temperature is the high heat capacity of liquid water. *Heat capacity* is the amount of heat energy that is required to raise the temperature of a substance 1°C.

Raising the temperature of liquid water takes a relatively large amount of heat. The temperature of a given quantity of water is raised only 1°C by an amount of heat that would increase the temperature of the same quantity of ethyl alcohol by 2°C, or of chloroform by 4°C. This phenomenon contributes to the surprising constancy of the temperature of the oceans and other large bodies of water through the seasons of the year. The temperature changes in coastal land masses are also moderated by large bodies of water. Indeed, water helps minimize variations in atmospheric temperature throughout the planet.

Cohesion and Surface Tension

In liquid water, the molecules are free to move about. The hydrogen bonds between the water molecules continually form and break. In other words, liquid water has a dynamic structure. On the average, every water molecule forms 3.4 hydrogen bonds with other water molecules. This number represents fewer bonds than exist in ice, but it is still a high number.

These hydrogen bonds explain the cohesive strength of water. The *cohesive strength* of water is what permits narrow columns of water to stretch from the roots to the leaves of trees more than 100 meters high. When water evaporates from leaves, the entire column moves upward in response to the pull of the molecules at the top.

Water also has a high *surface tension*, which means that the surface of water exposed to the air is difficult to puncture. The water molecules in this surface layer are hydrogen-bonded to other water molecules below. The surface tension of water permits a container to be filled slightly above its rim without overflowing, and it permits small animals to walk on the surface of water.

Evaporation and Cooling

Water has a high *heat of vaporization*; which means a lot of heat is required to change water from its liquid state to its gaseous state (the process of evaporation). This heat is absorbed from the environment in contact with the water. Evaporation thus has a cooling effect on the environment-whether a leaf, a forest, or an entire land mass. This effect explains why sweating cools the human body: As the sweat evaporates off the skin, it takes with it some of the adjacent body heat.

Water Molecules Sometimes Form Ions

The water molecule has a slight but significant tendency to come apart into a hydroxide ion (OH^-) and a hydrogen ion (a proton, H^+). Actually, *two* water molecules participate in ionization.

The hydronium ion is in effect a. hydrogen ion bound to a water molecule. For simplicity, biochemists tend to use a modified representation of the ionization of water:

$$H_2O \rightarrow H^+ + OH^-$$

Even though only about one water molecule in 500 million is ionized at any given time, the transformation is significant because H^+ and OH^- ions participate in many important biochemical reactions.

Acids, Bases, and the pH Scale

The ionization of water is very important for all living creatures. This fact may seem surprising, since the ionization is so slightonly one ionization out of 5×10^8 water molecules. But we are less surprised if we focus on the abundance of water in living systems and the reactive nature of H^+ produced by that ionization. Remember that H^+ is a proton, a tiny bit of charged matter-smaller than any atom or molecule in the cell.

Usually it attaches to another water molecule, but in the cell it can attach to other molecules and substantially change their properties. Because acids and bases donate and accept H^+, they can profoundly alter the water environment in which the chemical reactions and processes of life take place.

In this section we'll examine acids, bases, and their measurement using the pH scale. We'll close by looking at how buffers limit the changes in pH.

Acids Donate H^+, Bases Accept H^+

In pure water, the concentration of hydrogen ions exactly equals that of hydroxide ions (OH^-), and this "solution" is said to be *neutral.* Now suppose we add some HCl (hydrochloric acid). As it dissolves, the HCl ionizes, releasing H^+ and Cl^- ions:

$$HCl \rightarrow H^+ + Cl^-$$

Now there are more H^+ than OH^- ions. Such a solution is *acidic*. A *basic*, or alkaline, solution is one in which there are more OH^- than H^+ ions. A basic solution can be made from water by adding, for example, sodium hydroxide (NaOH), which ionizes to yield OH^- and Na^+ ions, thus making the concentration of OH^- ions greater than that of H^+ ions.

$$NaOH \rightarrow Na^+ + OH$$

An acid is any compound that can *release* H^+ ions in solution. HCl is an acid, as is H_2SO_4 (sulfuric acid). One molecule of sulfuric acid may ionize to yield two H+ ions and one $S04^-$ ion. Biological compounds such as acetic acid and pyruvic acid, which contain -COOH (the carboxyl group) are also acids, because —COOH → -COO⁻ + H⁺.

Bases are compounds that can *accept* H^+ ions. These include the bicarbonate ion (HCO_3^-), which can accept a H^+ ion and become carbonic acid (H_2CO_3); ammonia (NH_3), which can accept a H+ ion and become an ammonium ion (NH_4^+); and many others.

Note that although —COOH is an acid, —COO^- is a base, because -COO^- + H^+ → —COOH. Acids and bases exist as pairs, such as —COOH and —COO^-, because any acid becomes a base when it releases a proton, and any base becomes an acid when it gains a proton.

You may have noticed that the two reactions just discussed are the opposites of each other. The reaction that yields —COO^- and H^+ is reversible and may be expressed as

$$—COOH \rightleftharpoons —COO^- + H^+$$

A *reversible reaction* is one that can proceed in either direction-left to right or right to left-depending on the relative starting concentrations of reacting substances and products. In principle, all chemical reactions are reversible.

pH is the Measure of Hydrogen Ion Concentration

The terms "acidic" and "basic" refer only to *solutions*. How acidic or basic a solution is depends on the relative concentrations of H^+ and OH^- ions in it. "Acid" and "base" refer to *compounds* and *ions*. A compound or ion that is an acid can donate H^+ ions; one that is a base can accept H^+ ions.

How do we specify how acidic or basic a solution is? First, let's look at the H^+ ion concentrations of a few contrasting solutions. In pure water, the H+ concentration is 10^{-7} *M*. In 1 M hydrochloric acid, the H^+ concentration is 1 *M;* and in 1 *M* sodium hydroxide, the H^+ concentration is 10^{-14} *M*. Because its values range so widely-from more than 1.0 *M* to less than 10^{-14} M the H^+ concentration itself is an inconvenient quantity. It is easier to work with the logarithm of the concentration, because logarithms compress this range. We indicate how acidic or basic a solution is by its pH (a term derived from "potential of Hydrogen"). The pH value is defined as the negative logarithm of the hydrogen ion concentration in moles per liter (molar concentration). In chemical notation, molar concentration is often indicated by putting brackets around the symbol for a substance; thus $[H^+]$ stands for the molar concentration of H^+. The equation for pH is

$$pH = -\log_{10}[H^+]$$

Since the H^+ concentration of pure water is 10^{-1} *M,* its pH is -log(10^{-7}) = -(-7), or 7. A smaller negative logarithm means a larger number. In practical terms, a lower pH means a higher H^+ concentration, or greater acidity. In 1 *M* HCl, the H^+ concentration is 1 *M,* so the pH is the negative logarithm of 1 (–log 10^0), or 0. The pH of 1 *M* NaOH is the negative logarithm of 10^{-14}, or 14. A solution with a pH of less than 7 is acidic: It contains more H^+ ions than OH^- ions. A solution with a pH of 7 is neutral. And a solution with a pH value greater than 7 is basic. Because the pH scale is logarithmic, the values are exponential: A solution with a pH of 5 is 10 times more acidic than one with a pH of 6 (it has ten times as great a concentration of H^+); a solution with a pH of 4 is 100 times more acidic than one with a pH of 6.

Buffers Minimize pH Changes

An organism must control the chemistry of its cellsin particular, the pH of the separate compartments within cells. Animals must also control the pH of their blood. The normal pH of human blood is 7.4, and deviations of even a few tenths of a pH unit can be fatal. The control of pH is made possible in part by *buffers*, systems that maintain a relatively constant pH even when substantial amounts of acid or base are added. A buffer is a mixture of an acid that does not ionize completely in water and its corresponding base-for example, carbonic acid (H_2CO_3) and bicarbonate ions (HCO_3^-). If acid is added to this buffer, not all the H^+ ions from the acid stay in solution. Instead, many of the added H^+ ions combine with bicarbonate ions to produce more carbonic acid, thus using up some of the H+ ions in the solution and decreasing the acidifying effect of the added acid:

$$HCO_3 + H^+ \rightleftharpoons H_2CO_3$$

If base is added, the reaction reverses. Some of the carbonic acid ionizes to produce bicarbonate ions and more H^+, which counteracts some of the added base.

In this way, the buffer minimizes the effects of added acid or base on pH. A given amount of acid or base causes a smaller change in pH in a buffered solution than in an unbuffered one. Buffers illustrate the reversibility of chemical reactions: Addition of acid drives the reaction in one direction; addition of base drives it in the other.

Properties of Molecules

Molecules vary in size. Some are small, such as H_2 and CH_4. Others are larger, such as a molecule of table sugar (sucrose), which has 45 atoms. Still other molecules, such as proteins, are *gigantic*, sometimes containing tens of thousands of *atoms bonded* together in specific ways. Whether large, *medium*, or small, most of the molecules in living systems contain carbon atoms and are thus referred to as *organic molecules*. Most organic molecules include hydrogen and oxygen atoms, and many also include nitrogen and phosphorus.

All molecules have a specific three-dimensional shape. For example, the orientation of the bonding orbitals around the carbon atom gives the molecule of methane (CH_4) the shape of a regular tetrahedron, while in carbon dioxide (CO_2) the three atoms are in line with one another. Larger molecules have specific complex shapes that result from the number and kinds of atoms present and the ways in which these atoms are linked together. Some large molecules have compact ball-like shapes. Others are long, thin, ropelike structures. The shapes relate to the roles these molecules play in living cells.

In addition to size and shape, molecules have certain properties that characterize them and determine their biological roles. Chemists can use the characteristics of composition, structure (three-dimensional shape), reactivity, and solubility to distinguish a sample of one pure compound from another. For example, a compound such as sugar is soluble in water, which means that the solid disperses to form a uniform homogeneous mixture, but the same compound is insoluble in oil (the solid sugar remains a solid), and the resulting mixture is heterogeneous. Another substance, such as butane, is insoluble in water but highly soluble in oil. These solubility differences are due to the different atoms present and to their arrangement in these two kinds of molecules.

The presence of polar or charged sites on a molecule plays important roles in determining the molecule's solubility in water. Such sites can also determine the kinds of chemical reactions in which the molecule participates. The sizes, shapes, solubilities, and reactivities of molecules are significant to understanding the structures and operations of cells and organisms. Certain groups of atoms found together in a variety of molecules simplify our understanding of the reactions that molecules undergo.

Functional Groups give Specific Properties to Molecules

On the basis of atomic composition, structure, and reactivity, we can distinguish families of molecules from other families. Each member within a family of compounds has some characteristics in common with the other members of that family but differs from them in other characteristics. For example, organic acids are a family of carbon compounds that are all acidic but differ in other ways. They all contain a characteristic group of atoms called the *carboxyl group*,

$$\begin{array}{c} O \\ || \\ \text{—C—OH, or COOH} \end{array}$$

that is the source of the H^+ that defines an acid:

$$\begin{array}{c} O \\ || \\ \text{—C—OH} \end{array} \rightarrow \begin{array}{c} O \\ || \\ \text{—C—}O^- \end{array} + H^+$$

The carboxyl group is a functional group. Functional groups are groups of atoms that are part of a larger molecule and have particular reactive characteristics. The same functional group may be part of very different molecules. In addition to the carboxyl group, you will encounter several other functional groups in your study of biology.

Several classes of biologically important compounds are defined by the functional groups they contain. When the functional group is a hydroxyl group (—OH), the product is an alcohol. Perhaps the most familiar alcohol is ethanol (also called ethyl alcohol, CH_3CH_2OH). Small alcohols like ethanol are soluble in water because of the hydrogen bonding possible between the polar hydroxyl group and water molecules, but larger alcohols are not soluble in water because of their long hydrocarbon chains.

Sugars contain both hydroxyl and carbonyl groups. The *carbonyl group* has a central carbon atom with a double bond to an oxygen atom (C=O). If one of the other two bonds of the carbon atom in a carbonyl group is attached to a hydrogen atom, the compound is an *aldehyde*.

Carboxyl groups are found in organic acids. Some organic bases, called amines, possess an *amino group* ($—NH_2$), which has a tendency to react with H^+ to produce the positively charged ammonium group ($—NH_3^+$). This H^+-accepting characteristic accounts for the classification of amines as bases.

Amino acids are important compounds that possess both a carboxyl group and an amino group attached to the same carbon atom, the a (alpha) carbon. Also attached to the a carbon atom are a hydrogen atom and a side chain designated by the letter R. Different side chains have different chemical compositions, structures, and properties. Each of the 20 amino acids found in proteins has a different side chain that gives it its distinctive chemical properties. Because they possess both carboxyl and amino groups, amino acids are simultaneously acids and bases. At the pH values commonly found in cells, both the carboxyl and the amino groups are ionized: The carboxyl group has lost α proton, and the amino group has gained one.

Two other functional groups deserve our attention: the sulfhydryl group and the phosphate group. The *sulfhydryl group* (—SH) is an important constituent of the side chains of two amino acids. When two of these sulfhydryl groups react, the two hydrogen atoms are lost and a covalent bond, called a disulfide bond, is formed between the sulfur atoms (—S—S—). As we'll see in the next chapter, disulfide bonds help maintain the three-dimensional structure of proteins that is required for normal protein functioning.

The *phosphate group* ($—OPO_3^-$) is found on many different kinds of molecules. It plays important structural roles in DNA and RNA and participates in reactions that transfer energy. The removal (hydrolysis) of phosphate groups from some molecules releases energy that can be used to fuel other energy-requiring reactions.

Isomers have Different Arrangements of the Same Atoms

Isomers are compounds that have the same chemical formula but different arrangements of the atoms. (The prefix "iso-" means "same and is encountered in many technical terms.) Of the different kinds of isomers, we will consider two: structural isomers and optical isomers.

Structural isomers are isomers that differ in how the atoms are joined together. Consider two simple molecules, each composed of four carbon and ten hydrogen atoms bonded covalently, with the formula C_4H_{10}. These atoms can be linked in two alternative ways in which carbon can form four bonds and hydrogen one bond. These two forms are called butane and isobutane. Their different bonding relationships are distinguished in structural formulas, and they have different chemical properties.

```
        H  H                    CH3
        |  |                     |
   H3C—C—C—CH3              H3C—C—CH3
        |  |                     |
        H  H                     H
      Butane                 Isobutane
```

Many molecules of biological importance, particularly the sugars and amino acids, have optical *isomers*. Optical isomers (also called enantiomers) are related to each other in the way an object is related to its mirror image. Optical isomers occur whenever a carbon atom has four *different* atoms or groups attached to it. These instances allow two different ways of making the attachments, each the mirror image of the other. Such a carbon atom is an asymmetric carbon, and the pair of compounds are optical isomers of each other. Your right and left hands are optical isomers. Just as a glove is specific for a particular hand, so some biochemical molecules can interact with a specific optical isomer of a compound but are unable to "fit" the other.

The a carbon in an amino acid is an asymmetric carbon because it is bonded to four different groups. Therefore, amino acids exist in two isomeric forms, called D-amino acids and 1.-amino acids. "D" and "L" are abbreviations for right and left, respectively. Only L-amino acids are commonly found in most proteins of living things.

The compounds discussed in this chapter include some of the more common ones found in organisms.

Between these small molecules and the world of the living stands another level, that of the giant macromolecules. These huge molecules—the proteins, lipids, carbohydrates, and nucleic acids—are the subject of the next chapter.

6

MACROMOLECULES

The lives of cells are dances and dramas with tens of thousands of different kinds of molecules. Their dramatic choreography is what scientists reveal as they investigate the molecules of living systems, their physical properties, and their chemical reactions. What molecules are present? What are their chemical structures and properties? And what biological functions do these molecules perform? In different ways, we will be concerned with these questions throughout much of this book.

In this chapter, we'll look at gigantic molecules called macromolecules (*macro-*, "large") and the subunits from which they are constructed. Macromolecules may contain hundreds or thousands of atoms and have very large molecular weights. For example, human hemoglobin has a molecular weight of 64,500 and contains more than 6,000 atoms. The molecular complexity of such a structure can be understood in terms of the structures and properties of its subunits, which are fewer in number and easier to understand. Macromolecules are chains of small, individual units called monomers (*mono-*, "one"; *-mer,* "unit") covalently bonded together to form a polymer (*poly-*, "many").

In living systems there are three types of macromolecules: polysaccharides, proteins, and nucleic acids. *Polysaccharides* are constructed from sugar monomers such as glucose; *proteins* are constructed from amino acids; and *nucleic acids* are composed of nucleotides. Another important group of molecules that can form large structures is the *lipids*. But because the individual components (the lipids) do not covalently bond together, the resulting large structures are not true macromolecules.

The monomers that form the basis of polysaccharides, proteins, and nucleic acids are identical in different species. For example, a molecule of glucose in human blood is identical to a molecule of glucose from a reptile or a cabbage plant. The same group of 20 different amino acids is used to construct proteins for all living things--from bacteria to birds, bats, and humans. However, this cannot always be said of the larger molecular forms built from these monomers.

Proteins and nucleic acids in particular are informational macromolecules that are different in different species. For example, human hemoglobin is different from the hemoglobin found in fish or birds. No animal acquires its macromolecules directly from its food. Instead, it uses the subcomponents of its food to construct new macromolecules suited to its unique needs. For example, we eat proteins constructed by other animals and plants, but we break these proteins down into their amino acids and then reassemble the amino acids into the chemically different proteins of our own bodies. This process is like picking up Lego toys that somebody else has made, taking them apart, and putting the parts together again to make the toys we want.

Before we turn to the main focus of this chapter—the macromolecules: carbohydrates, proteins, and nucleic acids—let's first take a look at lipids and the large macromolecule-like structures that they form.

Lipids: Water-Insoluble Molecules

Lipids are a chemically diverse group of hydrocarbons. The property they all share is an insolubility in water that is due to the presence of many nonpolar covalent bonds. Nonpolar molecules can associate together and form massive structures, but these structures are not considered macromolecules, because the bonds between the separate molecules are not covalent bonds. These nonpolar hydrocarbon molecules are literally pushed together by surrounding water molecules, which are not attracted to the nonpolar substance. When the nonpolar molecules are sufficiently close together, weak but additive van der Waals forces hold them together. Although insoluble in water, lipids are soluble in other lipids or in nonpolar solvents such as ether or benzene.

In addition to their diverse chemical structures, lipids have many different biological roles. Some of them store energy (the fats and oils). Others play important structural roles in cell membranes (phospholipids). The carotenoids help plants capture light energy, and

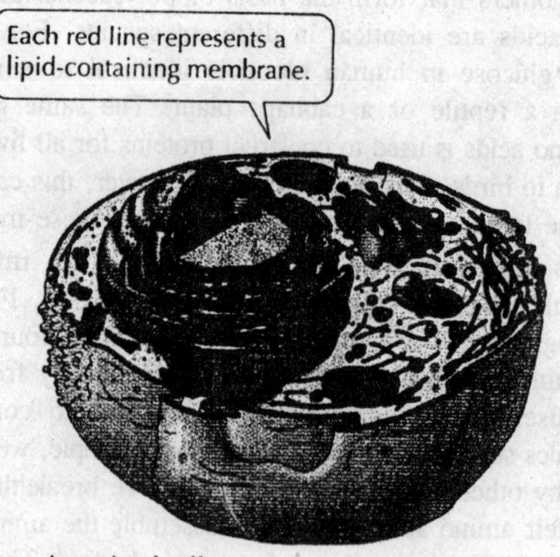

An animal cell

Fig. 6.1. Lipids from cellular membranes.

the steroids and some lipids play regulatory roles. We'll begin our discussion by considering the structures and properties of the fats and oils.

Fats and Oils Store Energy

Chemically, fats and oils are triglycerides, also known as *simple lipids*. Triglycerides that are solid at room temperature (20°C) are called fats; those that are liquid at room temperature are called *oils*.

Triglycerides are composed of two types of building blocks: fatty acids and glycerol. *Glycerol* is a small molecule with three hydroxyl (—OH) groups. Fatty *acids* have long nonpolar hydrocarbon tails and a polar carboxyl functional group (—COOH). Fatty acids can be of different lengths (usually 16 to 20 carbon atoms) and may be either Theturated or unsaturated.

In *saturated fatty acids*, all the bonds between the carbon atoms in the hydrocarbon chain are single bonds-there are no double bonds. That is, all the bonds are *saturated* with hydrogen atoms. Saturated fatty acids include palmitic acid, which has 16 carbon atoms, and stearic acid, which has 18. These molecules are relatively rigid and straight, and they pack together tightly, like pencils in a box.

In *unsaturated fatty acids*, the hydrocarbon chain contains one or more double bonds. Oleic acid is a monounsaturated fatty acid that has 18 carbon atoms, and one double bond near the middle of the hydrocarbon chain causes a kink in the molecule. Fatty acids, such as

(*a*) Palmitic acid

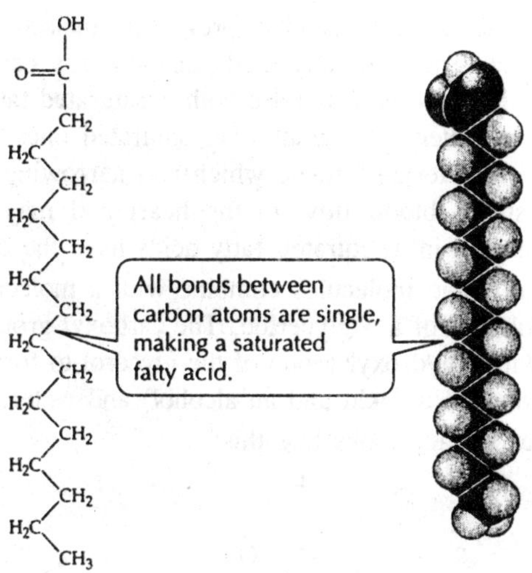

(*b*) Oleic acid

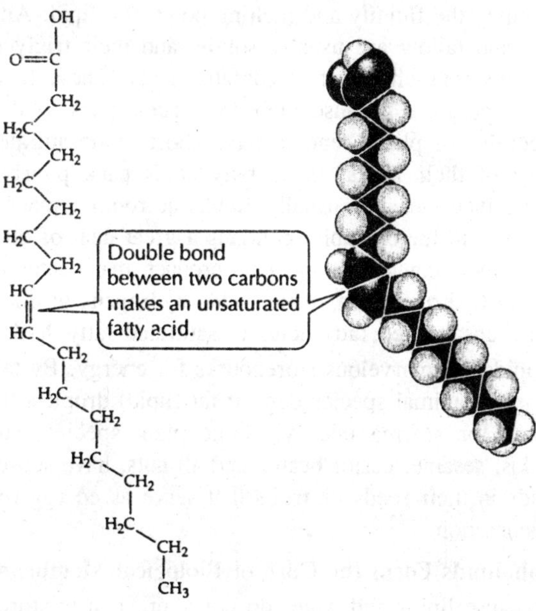

Fig. 6.2. Fatty acids.

linoleic acid, that have more than one double bond are polyunsaturated and have multiple kinks. These kinks prevent the molecules from packing together tightly. The two fatty acids that humans cannot synthesize and must obtain from their diet are both unsaturated fatty acids. Diets favouring unsaturated fatty acids over saturated ones tend to reduce the incidence of arteriosclerosis, which is a narrowing of the arterial wall that restricts blood flow to the heart and may lead to heart attacks. Diets rich in Theturated fatty acids have the opposite effect.

Three fatty acid molecules combine with a molecule of glycerol to form a molecule of a triglyceride. The carboxyl group of each fatty acid reacts with a hydroxyl group of the glycerol to form an *ester* (the reaction product of an acid and an alcohol) and water.

The ester linkage looks like this:

O
||
—C—O—C—

The three fatty acids in one triglyceride molecule need not all have the same length, nor do they all have to be either saturated or unsaturated. The kinks associated with double bonds are important in determining the fluidity and melting point of a lipid. Animal fats such as lard and tallow are usually solids, and their triglycerides tend to have many long-chain saturated fatty acids. These fats are solids at room temperature because their fatty acids pack well together. The triglycerides of plants tend to have short or unsaturated fatty acids. Because of their kinks, these fatty acids pack poorly together and these triglycerides are usually liquid at room temperature. Natural peanut butter, for example, contains a great deal of oil. However, to market a more convenient and solid product, manufacturers often subject the oil to hydrogenation, which adds hydrogens to double bonds and converts unsaturated fatty acids to saturated fatty acids.

Lipids are marvelous storehouses for energy. By taking in excess food, many animal species deposit fat (lipid) droplets in their cells as a means for storing energy. Some plant species, such as olives, avocados, sesame, castor beans, and all nuts, have substantial amounts of lipids in their seeds or fruits that serve as energy reserves for the next generation.

Phospholipids Form the Core of Biological Membranes

Because lipids and water do not interact, a mixture of water and lipids forms two distinct phases. Many biologically important substances—such as ions, sugars, and free amino acids—that are soluble in water are insoluble in lipids. These two properties—water solubility and lipid insolubility—have proved essential to distinguishing the interior of cells from their external environment.

Suppose that you must design water-filled compartments, separated from each other and from their environment by barriers that limit the passage of materials. Given the properties of lipids, a seemingly effective way to accomplish this task is to use membranes that contain a special class of lipid called the phospholipid. This is the system that has evolved in nature. Molecular traffic within an organism or into and out of its compartments is constrained by the properties of the lipid portion of the surrounding membrane. Compounds that dissolve readily in lipids can move rapidly through biological membranes, but compounds that are insoluble in lipids are prevented from passing through the membrane or must be transported across the membrane by specific proteins.

Like triglycerides, phospholipids have fatty acids bound to glycerol by ester linkages. In phospholipids, however, any one of several phosphate-containing compounds may replace one of the fatty acids. Many phospholipids are important constituents of biological membranes. If you think carefully about the structure and properties of phospholipids, you will find it easy to understand how they are oriented in membranes. The phosphate functional group has negative electric charges, so this portion is hydrophilic, attracting polar water molecules. But the two fatty acids are hydrophobic, so they are pushed together by water.

In a biological membrane, phospholipids line up in such a way that the nonpolar, hydrophobic "tails" pack tightly together to form the interior of the membrane, and the phosphate-containing "heads" face outward (some to one side of the membrane and some to the other), where they interact with water, which is excluded from the interior of the membrane. The phospholipids thus form a bilayer, a sheet two molecules thick. Because membranes are so important.

Because the word "lipid" defines compounds in terms of their solubility rather than their structural similarity, a great variety of different chemical structures are included as lipids. The next two lipid classes we'll discuss—the carotenoids and the steroids—have chemical structures very different from the structures of triglycerides and phospholipids and from the structures of each other.

Carotenoids Trap Light Energy

The *carotenoids* are a family of light-absorbing pigments found in plants and animals. Betacarotene (β-carotene) is one of the pigments that traps light energy in leaves during photosynthesis. It is the β-carotene in plants that senses light and causes their parts to grow toward or away from the light. In humans, a molecule of β-carotene can be broken down into two vitamin A molecules, from which we make the pigment rhodopsin, which is required for vision. Carotenoids are responsible for the colour of carrots, tomatoes, pumpkins, egg yolks, and butter.

Steroids are Signal Molecules

The *steroids* are a family of organic compounds whose multiple rings share carbons. Some steroids are important constituents of membranes. Others are hormones, chemical signals that carry messages from one part of the body to another. *Testosterone* and the *estrogens* are steroid hormones that regulate sexual development in vertebrates. *Cortisol* and related hormones play many regulatory roles in the digestion of carbohydrates and proteins, in the maintenance of salt balance and water balance, and in sexual development.

Cholesterol is synthesized in the liver and contributes to the structure of some cellular membranes. It is the starting material for making testosterone and other steroid hormones, as well as the bile salts that help break down dietary fats so that they can be digested.

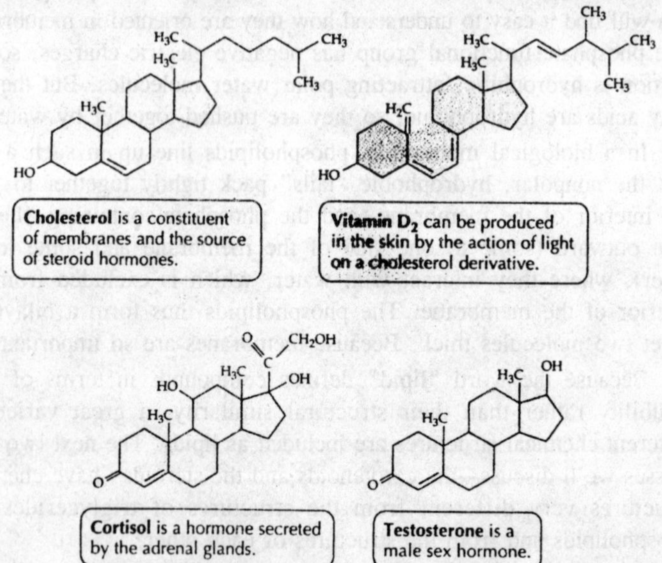

Fig. 6.3. All steroids have the same ring structures.

We absorb cholesterol from foods such as milk, butter, and animal fats. When we have too much cholesterol in our blood, it is deposited in our arteries (along with other substances), a condition that may lead to arteriosclerosis and heart attack.

Some Lipids are Vitamins

A large group of fat-soluble substances, including the carotenoids and steroids, are synthesized by covalent linking and chemical modification of isoprene to form a series of isoprene units:

$$CH_2 = \underset{}{\overset{\overset{\displaystyle CH_3}{|}}{C}}\text{—}CH = CH_2$$

The fat-soluble vitamins (A, D, E, and K) are formed in this manner by plants and bacteria. These substances are not synthesized by humans and must be acquired from dietary sources. *Vitamin A* is formed from β-carotene found in green and yellow vegetables. Among other roles, vitamin A is directly involved in the reception of light by our eyes. Deficiency in vitamin A leads to dry skin, eyes, and internal body surfaces; retarded growth and development; and night blindness, which is a diagnostic symptom for the deficiency. *Vitamin D* regulates the absorption of calcium from the intestines. It is necessary for the proper deposition of calcium in bones; a deficiency of vitamin D can lead to rickets, a bonesoftening disease.

Vitamin E is not a single vitamin, but a group of related lipids that seem to protect cells from damaging effects of oxidation-reduction reactions. These lipids appear to have an important role in preventing unhealthy changes in the double bonds in the unsaturated fatty acids of membrane phospholipids. Commercially, vitamin E is added to some foods to slow spoilage. *Vitamin K* is found in green leafy plants and is also synthesized by bacteria normally present in the human intestine. This vitamin is essential to the formation of blood clots. Predictably, a deficiency of vitamin K leads to slower clot formation and potentially fatal bleeding from a wound.

Because of their insolubility in water, in the body some lipids may require water-soluble carrier proteins in order to be transported in the blood, which is mostly water. Among the lipids, only triglycerides and phospholipids assemble into large structures in cells: triglycerides form droplets of stored fat, and phospholpids form the bilayers of membranes. But in spite of their size, these droplets and bilayers are not usually considered macromolecules.

Now that we have seen the macromolecule-like structures that lipids form, let's turn to the large structures that constitute true macromolecules.

Macromolecules: Giant Polymers

As noted earlier, *macromolecules* are giant *polymers* constructed by the covalent linking of smaller molecules called *monomers*. These monomers may or may not be identical, but they always have similar chemical structures. Molecules with molecular weights exceeding 1,000 are usually considered macromolecules, and the polysaccharides, proteins, and nucleic acids of living systems certainly fall into this category.

In the cell, each type of macromolecule performs some combination of a diversity of functions: energy storage, structural support, protection, catalysis, transport, defense, regulation, movement, and heredity. These roles are not necessarily exclusive. For example, both carbohydrates and proteins can play structural roles-supporting and protecting cells and organisms. However, only nucleic acids specialize in information and function as hereditary material, carrying both species and individual traits from generation to generation.

The functions of macromolecules are directly related to their shapes and the chemical properties of their monomers. Some macromolecules, such as catalytic and defensive proteins, fold into compact spherical forms with surface features that make them watersoluble and capable of intimate interaction with other molecules. Other proteins and carbohydrates form long, fibrous systems that provide strength and rigidity to cells and organisms. Still other long, thin assemblies of proteins can contract and cause movement.

Because macromolecules are so large, they contain many different functional groups. For example, a large protein may contain hydrophobic, polar, and charged groups. These groups give specific properties to local sites on a macromolecule. As we will see, this diversity of properties determines the shapes of macromolecules and their interactions with both other macromolecules and smaller molecules.

Macromolecules Form by Condensation Reactions

The polymers of living things are constructed by a series of reactions called *condensation reactions*, or dehydration hydration reactions (both words refer to the loss of water). Condensation reactions covalently bond monomers. Water is lost in the reaction:

$$A{-}H + B{-}OH \rightarrow A{-}B + H_2O$$

A—H is a molecule with a reactive hydrogen; B—OH is a molecule with a hydroxyl group. The products of the reaction are A—B and H_2O. The atoms that make up the water molecule are derived from the reactants: one hydrogen atom from one reactant, and an oxygen atom and the other hydrogen atom from the other reactant. Condensation reactions are not limited to polymer formation; for example, we encountered them in the linkage (esterification) of fatty acids to glycerol.

The condensation reactions that produce the different kinds of macromolecules differ in detail, but in all cases polymers will form only if energy is added to the system. In living systems, specific energy-rich molecules supply the energy, and there are additional steps to the reaction. The reverse of a condensation reaction is a *hydrolysis reaction* (*hydro-*, "water"; *-lysis,* "breakage"). These reactions digest polymers and produce monomers. Water reacts with the bonds that link the monomers together, and the products are free monomers. The elements of the reactant H_2O become part of the products. Hydrolysis reactions of all sorts are very important in cellular functioning and are not limited to the digestion of polymers.

Carbohydrates: Sugars and Sugar Polymers

Carbohydrates are a diverse group of compounds based on the general formula CH_2O. Some are relatively small, with molecular weights less than 100. Others are true macromolecules, with molecular weights of hundreds of thousands. There are four categories of biologically important carbohydrates. *Monosaccharides* (*mono-*, "one"; *saccharide,* "sugar") such as glucose or fructose—are the monomers out of which the larger forms are constructed. *Disaccharides* (*di-*, "two") consist of two monosaccharides. *Oligosaccharides* (*oligo-*, "several") have several monosaccharides (3 to 20). *Polysaccharides* (*poly-*, "many")-such as starch, glycogen, and cellulose—are composed of hundreds of thousands of glucose units.

The relative proportions of carbon, hydrogen, and oxygen indicated by the general formula for carbohydrates, CH_2O, is true for monosaccharides. However, for disaccharides, oligosaccharides, and polysaccharides, these proportions differ slightly from this general formula because two hydrogens and an oxygen are lost during the condensation reactions.

Monosaccharides are Simple Sugars

All living cells contain the monosaccharide glucose, whose formula is $C_6H_{12}O_6$. Green plants produce glucose by photosynthesis, and other

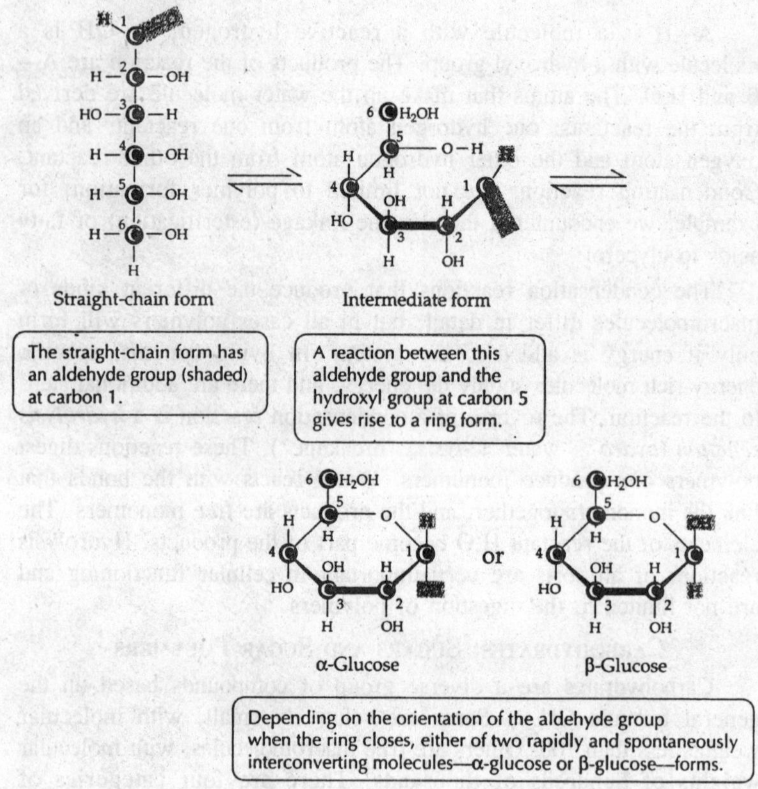

Fig. 6.4. Glucose: from one form to the other.

organisms acquire it directly or indirectly from plants. Cells use glucose as an energy source, changing it through a series of reactions that release stored energy and produce water and carbon dioxide.

Two forms of glucose, the straight chain and the ring, exist in equilibrium with each other when dissolved in water, but the ring form predominates (>99%). The two distinct ring forms (α and β-glucose) differ only in the placement of the —H and —OH attached to carbon 1. Most of the monosaccharides found in living systems belong to the D series of optical isomers. But there are structural isomers-composed of the same kinds and numbers of atoms, but with the atoms combined differently in each. All *hexoses* (*hex*-, "six"), a group of structural isomers, have the formula $C_6H_{12}0_6$. Included among the hexoses are fructose (so named because it was first found in fruits), mannose, and galactose.

Three-carbon sugar

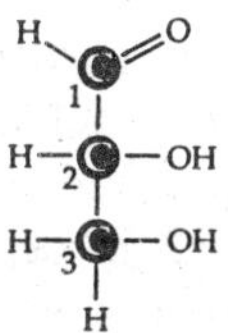

Glyceraldehyde is the smallest sugar and exists only as the straight-chain form.

Five-carbon sugars

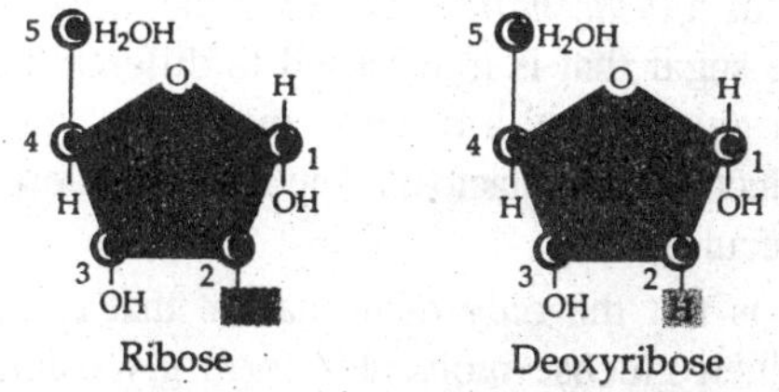

Ribose and deoxyribose each have five carbons, but very different chemical properties and biological roles.

Six-carbon sugars

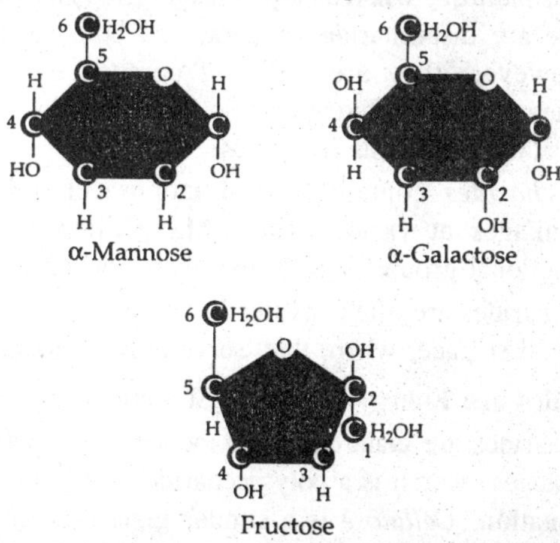

These hexoses are isomers. All have the formula $C_6H_{12}O_6$–but each has distinct chemical properties and biological roles.

Fig. 6.5. Monosaccharides are simple sugars.

Pentoses (*pent-*, "five") are five-carbon sugars. Some pentoses are found primarily in the cell walls of plants, as are several of the hexoses. Two pentoses are of particular importance: *Ribose* and *deoxyribose* form part of the backbones of RNA and of DNA, respectively. These two pentoses are not isomers; rather, one oxygen atom is missing from carbon 2 in deoxyribose (de, "absent"). The absence of this oxygen atom has enormous consequences for the functional distinction of RNA and DNA.

Glycosidic Linkages bond Monosaccharides Together

Monosaccharides are covalently bonded together by condensation reactions that form *glycosidic linkages*. Such a linkage between two simple sugars forms a *disaccharide*. For example, a molecule of *sucrose* (table sugar)—the sugar that is transported to different parts of plants—is formed from a glucose and a fructose molecule, while *lactose* (milk sugar) contains glucose and galactose. The disaccharide *maltose* contains two glucose molecules.

But maltose is not the only disaccharide that can be made from two glucoses. When glucose molecules form glycosidic linkages, the disaccharide product must be one of two types: α-linked or β-linked, depending on whether the molecule that bonds by its 1 carbon is α-glucose or β-glucose. An α linkage with carbon 4 of a second glucose molecule gives *maltose*, whereas a β linkage gives *cellobiose*. Maltose and cellobiose are disaccharide isomers, and both have the formula $C_{12}H_{22}O_{11}$. However, they are different compounds with different properties. They undergo different chemical reactions and are recognized by different catalytic proteins (enzymes).

Oligosaccharides contain several monosaccharides linked by glycosidic linkages at various sites. Many oligosaccharides gain additional functional groups, which give them special properties.

Oligosaccharides are often covalently bonded to proteins and lipids on the outer cell surface, where they serve as cell recognition signals.

Polysaccharides are Energy Stores or Structural Materials

Polysaccharides are giant chains of monosaccharides connected by glycosidic linkages. Starch is a polysaccharide of glucose with linkages in the a orientation. *Cellulose* is a similar giant polysaccharide made up solely of glucose, but its individual units are connected by β linkages instead of a linkages. Cellulose is the predominant component of plant cell walls and is by far the most abundant organic compound on this planet. Both starch and cellulose are composed of nothing but glucose, but their biological functions and chemical and physical properties are entirely different.

Starch can be more or less easily degraded by the actions of chemicals or catalytic proteins (enzymes). Cellulose, however, is very stable because of its β glycosidic linkages. Thus starch is a good storage form that can be easily degraded to supply glucose for energy producting reactions, while cellulose is an excellent structural component that can withstand harsh environmental conditions without changing.

Starch is not just one chemical substance, but large family of giant molecules of broa structure. All starches are large polymers of glucose with a linkages, but different starches can be distinguished by the amount of branching between carbons 1 and 6. Some starches are highly branched; others are not. Plant starches, called *amylose*, are not highly branched. The polysaccharide glycogen, which stores glucose in animal livers and muscles, is highly branched.

What do we mean when we say that starch and glycogen are storage compounds for energy? Very simply, these compounds can readily be hydrolyzed to yield glucose monomers. Glucose, in turn, can enter into a series of reactions that liberate its stored energy in forms that can be used for cellular activities. However, glucose can also serve the cell's need for raw materials—that is, carbon atoms. Glucose can undergo other chemical reactions that disassemble parts of the molecule and rearrange its carbon atoms to form the skeletons of other compounds needed by cells. Glycogen and starch are thus storage depots for carbon atoms as well as for energy.

Derivative Carbohydrates Contain other Elements

Sometimes carbohydrates are modified by chemical changes in their structure or by the addition of functional groups such as phosphate and amino groups, thus becoming *derivative carbohydrates*. For example, carbon 6 in glucose may be oxidized from —CH_2OH to a carboxyl group (—COOH), producing glucuronic acid. Or a phosphate group may be added to one or more of the —OH sites. Some of these *sugar phosphates*, such as fructose 1,6-bisphosphate, are important intermediates in cellular energy reactions. When an amino group is substituted for an —OH group, *amino sugars* such as glucosamine and galactosamine are produced. Galactosamine is a major component of cartilage, the material that forms caps on the ends of bones and stiffens the protruding parts of the ears and nose. A derivative of glucosamine produces the polymer *chitin*, which is the principal structural polysaccharide in the skeletons of insects, crabs, and lobsters, as well as in the cell walls of fungi. Fungi and insects (and their relatives) constitute more than 80 percent of the species ever described, and chitin is one of the most abundant substances on Earth.

Proteins: Amazing Polymers of Amino Acids

Proteins are an extraordinary group of macromolecules. Constructed of amino acids, proteins are fascinating because they have such intricate and diverse structures and because they perform so many different functions for cells. Among the cellular functions of macromolecules listed earlier, only energy storage and heredity are not functions of proteins. Proteins are involved in structural support, protection, catalysis, transport, defense, regulation, and movement. Of particular importance are the catalytic proteins, called *enzymes*, that increase the rates of chemical reactions in cells. In general, each chemical reaction requires a different enzyme, because proteins show great specificity for the smaller molecules with which they will interact.

Proteins range in size from the small RNA-digesting enzyme *ribonuclease A*, which has a molecular weight of 5,733 and 51 amino acid residues, to gigantic molecules such as the cholesterol transport protein *apolipoprotein B*, which has a molecular weight of 513,000 and 4,636 amino acid residues. (The word "residue" is used for a monomer when it is part of a polymer.) Each of these proteins consists of one chain of amino acids folded into a specific three-dimensional shape that is required for protein function. Some proteins have more than one polymer chain. For example, the oxygen-carrying protein *hemoglobin* has four chains that are folded separately and associate together to make the functional protein. The largest protein complex known has more than 40 separate chains of amino acids.

All of these different proteins have a characteristic amino acid composition, but every protein contains neither all 20 amino acids nor an equal number of different amino acids. Nor is there a simple regular sequence in which the amino acids are linked. The diversity in amino acid content and sequence is the source of the diversity in protein structures and functions. In some proteins, different kinds of chemical substances, called *prosthetic groups*, may be attached to the protein. These prosthetic groups include carbohydrates, lipids, phosphate groups, the iron-containing heme group, and metal ions such as copper and zinc. Whether they are small or large, and whether or not they have prosthetic groups, there is nothing "casual" about the structures of proteins. Proteins have specific three-dimensional shapes that are necessary for their specific functions.

To understand this stunning variety of functions, we must explore protein structure. First, we will examine the properties of the 20 amino acids and the characteristics of how they link to form proteins. Then we will systematically examine the four levels of protein structure and look at how a linear chain of amino acids is consistently folded into a compact three-dimensional shape.

Proteins are Composed of Amino Acids

The 20 different amino acids commonly found in proteins show a wide variety of properties. We considered the structure of amino acids and identified four different groups attached to a central carbon atom: a hydrogen atom, an amino group, a carboxyl group, and a side chain, or R group. Since amino acids are linked together by reactions between their amino and carboxyl groups, these groups are not exposed and do not give the protein its distinguishing properties and functional specificity-the side chains fill this role.

The *side chains* of amino acids are a protein's reactive groups and show a wide variety of chemical properties. Side chains control the function of a protein and contribute to its structure. The sequence of side chains determines how a protein folds into a three-dimensional shape (which we will discuss shortly). Although side chains are very important, they are often omitted from diagrams of proteins when the focus is on other aspects of structure. In these cases, the side chains are identified by the letter R (for "residue") and are sometimes called R groups.

One useful classification of amino acids is based on whether their side chains are electrically charged (+1, -1), polar (δ+, δ-), or nonpolar and hydrophobic. The five amino acids that have electrically charged side chains attract water and oppositely charged ions of all sorts. The four amino acids that have polar side chains tend to form weak hydrogen bonds with water and with other polar or charged substances. Eight amino acids have side chains that are nonpolar hydrocarbons or very slightly modified hydrocarbons. In the watery environment of the cell, the hydrophobic side chains may cluster together.

Three amino acids—cysteine, glycine, and proline—are special cases, although their side chains are generally hydrophobic. Two *cysteine* side chains, which have terminal —SH groups, can react to form a covalent bond in a *disulfide bridge* (—S—S—). Hydrogen bonds and disulfide bridges help determine how a protein chain folds. When cysteine is not part of a disulfide bridge, its side chain is very hydrophobic. The *glycine* side chain consists of a single hydrogen atom; thus glycines may fit into tight corners in the interior of a protein molecule, where a larger side chain could not fit. *Proline* differs from other amino acids because it possesses a modified amino group.

Peptide Linkages Covalently Bond Amino Acids Together

When amino acids polymerize, the carboxyl group of one amino acid reacts with the amino group of another, undergoing a condensation reaction that forms a peptide linkage. Gives a simplified description of the reaction. (In living cells, other molecules must activate the reactants in order for this reaction to proceed, and there are intermediate steps.) A linear polymer of amino acids connected by peptide linkages is a *polypeptide*. A *protein* is made up of one or more polypeptides.

At one end of the polypeptide is a free amino group. This end is the N terminus, named for the nitrogen atom in the amino group. At the other end of the polypeptide—the C terminus—is a free carboxyl group. The other amino and carboxyl groups are bound in peptide linkages. Thus a protein has direction. For example, the dipeptide glycine-alanine, in which glycine has the free amino group, differs from alanine-glycine, in which alanine has the free amino group. In cells, the synthesis of polypeptide chains begins with the N terminus.

In the peptide linkage, the C=O oxygen carries a slight negative charge (δ-), whereas the N—H hydrogen is slightly positive (δ+). This asymmetry of charge favours hydrogen bonding within the protein molecule itself and with other molecules, contributing to both the structure and the function of many proteins.

Primary Structure of a Protein is its Amino Acid Sequence

Protein structure is elegant and complex—so complex that it is described as consisting of four different levels: *primary*, *secondary*, *tertiary*, and *quaternary*. However, proteins are always linear chains; there are no branches. The precise sequence of amino acids in a polypeptide constitutes the primary *structure* of a protein. The *peptide backbone* of this primary structure consists of a repeating sequence of three atoms (—N—C—C—) from the amino group, the central carbon, and the carboxyl group of each amino acid.

In cells, the primary structure of a protein is dictated by the precise sequence of nucleotides in a linear segment of a DNA molecule. The elucidation of this relationship between DNA primary structure and protein primary structure was one of the triumphs of molecular biology.

The theoretical number of different proteins is enormous. Since there are 20 different amino acids, there are 20 × 20 = 400 distinct dipeptides, and 20 × 20 × 20 = 8,000 different tripeptides. Imagine this process of multiplying by 20 extended to a protein made up of 100 amino acids (which is considered a small protein): There could be 20^{100} of these small proteins, each with its,own distinctive primary structure. How large is the number 20^{100}? In the entire universe there aren't that many electrons!

At the higher levels of protein structure, local coiling and folding give the final functional shape of the molecule, but all of these levels derive from the primary structure. The different properties associated with a precise sequence of amino acids determine how the protein can twist and fold. By twisting and folding, each protein adopts a specific stable structure that distinguishes it from every other protein.

Secondary Structure of a Protein Requires Hydrogen Bonding

Although the primary structure of each protein is unique, the secondary structure of many different proteins may be the same. A protein's *secondary structure* consists of regular, repeated patterns in different regions of a polypeptide chain. One type of secondary structure, the *α helix* (alpha helix), is a right-handed coil that is "threaded" in the same direction as a standard wood screw. The amino acid side chains extend outward from the peptide backbone of the helix.

This helical structure of a polypeptide chain results from hydrogen bonds between elements of the peptide bonds that are distributed along the back bone of the chain. Hydrogen bonds form between the slightly positive hydrogen of the N—H of one peptide bond and the slightly negative oxygen of the C=O of another peptide bond. When this pattern of hydrogen bonding is established repeatedly over a segment of the protein, it stabilizes the twisted form, resulting in an a helix. However, the ability of a protein to form an a helix depends on its primary structure. Amino acids with larger side chains that distort the coil or otherwise prevent the formation of the necessary hydrogen bonds will keep the a helix from forming.

Alpha helical secondary structure is particularly evident in the fibrous structural proteins called *keratins*. Keratins constitute many of the protective materials found in mammals and birds, such as fingernails and claws, skin, hair, wool, and feathers. Hair can be stretched because this stretching requires that only hydrogen bonds in an a helix, and not covalent bonds, be broken; when the tension on the hair is released, both the helix and the hydrogen bonds reform.

Another type of secondary structure, the β pleated sheet, is found in the protein silk. In silk, rather than being coiled, the protein chains are almost completely extended and lie next to one another, stabilized by hydrogen bonds between the elements of the peptide linkages. The R pleated sheet may be found between separate polypeptide chains, as in silk, or between different regions of the same polypeptide that is bent back on itself. Many enzymes contain regions of a helix and of β pleated sheet in the same polypeptide chain.

Tertiary Structure of a Protein is Formed by Bending and Folding

In most proteins, to establish the compact structure the polypeptide chain must be bent at specific sites and folded back and forth. The resulting overall shape of a protein is its tertiary structure. Although the a helices and β pleated sheets contribute to the tertiary structure, more frequently only limited portions of the molecule have these secondary structures, and large regions consist of structures unique to a particular protein.

A complete description of the tertiary structure specifies the location of every atom in the molecule in three-dimensional space, in relation to all the other atoms. Bear in mind that this tertiary structure and the secondary structure derive from the protein's primary structure. If lysozyme is heated carefully, causing only the tertiary structure to break down, the protein will return to its normal tertiary structure when it cools. The only information needed to specify the unique shape of the lysozyme molecule is the information contained in its primary structure.

Hemoglobin is the protein that transports oxygen (O_2) from the lungs to the tissues. In muscle tissue, hemoglobin delivers O_2 to myoglobin, a smaller but similar protein, for storage. The hemoglobin molecule consists of four similar polypeptide chains that have tertiary structures made up almost entirely of a helices but that lack any disulfide bridges. The helical segments bend and fold against each other, forming a pocket that encloses a *heme group*, an ironcontaining prosthetic group that binds O_2. The stabilizing interaction of hydrophobic side chains on the inner sides of the a helices helps ensure that the helices fold against one another correctly as each polypeptide assumes its tertiary structure.

Quaternary Structure of a Protein Consists of Subunits

As we mentioned earlier, some proteins have two or more polypeptide chains folded into their own unique tertiary structures.

Each of these folded polypeptides is considered a protein subunit. *Quaternary structure* results from the ways in which these protein subunits fit together and interact; it can be illustrated by hemoglobin. Hydrophobic interactions, hydrogen bonds, and ionic bonds all help hold the four subunits together to form the functional hemoglobin molecule. As the hemoglobin molecule takes up one O_2 molecule, the four subunits shift their relative positions slightly, changing the quaternary structure. Ionic bonds are broken, exposing buried side chains that enhance the binding of additional O_2 molecules.

Each subunit of hemoglobin is folded like a myoglobin molecule, suggesting that both hemoglobin and myoglobin are evolutionary descendants of the same oxygen-binding ancestral protein. But on the surfaces where its subunits come in contact with each otherregions that on myoglobin are exposed to aqueous surroundings and are hydrophilic-hemoglobin has hydrophobic side chains. Again, the chemical nature of side chains on individual amino acids determines how the molecule folds and packs in three dimensions.

Molecular Chaperones Help Shape Proteins

The primary structure of a protein constrains the secondary, tertiary, and quaternary structures (if subunits exist). By determining the primary structure, DNA also determines the higher levels of structure. However, other factors also affect the tertiary structure that is required for proper protein function.

Elevated temperatures, pH changes, or altered salt concentrations can cause a protein to adopt a different, biologically inactive tertiary structure. Biological function depends on a specific three-dimensional structure. Increased temperature causes more rapid molecular movement and thus can break weak hydrogen bonds and hydrophobic interactions. Altered pH can change the pattern of ionization of carboxyl and amino groups in the side chains of amino acids, thus disrupting the pattern of ionic attractions and repulsions that contribute to normal tertiary structure.

The loss of appropriate tertiary structure is called *denaturation*, and it is always accompanied by a loss of the normal biological function of the protein. Denaturation can be caused by heat or high concentrations of polar substances such as urea that disrupt the hydrogen bonding that is crucial to protein structure. Nonpolar solvents may also disrupt normal structure. Usually denaturation is irreversible, particularly if many denatured proteins interact in random nonspecific ways. However, in some cases denaturation is reversible, and upon return to normal environmental conditions, the protein may return to its active form, as does lysozyme. This return is called *renaturation*.

How can such a complicated molecule spontaneously fold into its normal tertiary structure? Biologists are just beginning to be able to predict how a protein will fold, given its primary structure. The study of protein folding has a long way to go. The situation in the living cell is complex-even more complex than we first imagined. The protein doesn't form all at once. In a human cell it may take minutes from the beginning of synthesis until the entire protein squeezes into the compartment where it folds-and things are very crowded in that compartment. There, other proteins present inappropriate potential partners for interaction (in the form of amino acid side chains).

A special group of proteins, called the *chaperone proteins*, at least in part help prevent newly formed proteins from reacting inappropriately. The chaperones attach to a new protein while it is forming and protect it from interactions with other proteins until the new molecule has its correct shape. Most chaperone proteins can act on many different forming proteins. Some act on only a few, and others may act on only a single, specific protein.

What if the chaperone proteins fail to do their job, or if for other reasons a protein folds abnormally? Evidence suggests that abnormal protein folding underlies certain infectious diseases, including mad cow disease and, in humans, Creutzfeldt-Jakob disease. Alzheimer's disease may also result from protein misfolding.

Nucleic Acids: Informational Macromolecules

The nucleic acids are linear polymers specialized for the storage, transmission, and use of information. There are two types of nucleic acids: DNA (deoxyribonucleic acid) and RNA (ribonucleic acid). DNA molecules are giant polymers that encode hereditary information and pass it from generation to generation (through reproduction). The information encoded in DNA is also used to make specific proteins through the intermediate RNA. RNA molecules of various types copy the information in segments of DNA to specify the sequence of amino acids in proteins. Information flows from DNA to DNA in reproduction. But for nonreproductive activities of the cell, information flows from DNA to RNA to proteins, which ultimately carry out these functions. What compositions, structures, and properties of nucleic acids permit them to play these fundamental roles in living systems?

Nucleic Acids have Characteristic Structures and Properties

Nucleic acids are composed of monomers called *nucleotides*, each of which consists of a pentose sugar, a phosphate group, and a nitrogen-containing base-either a pyrimidine or a purine. Molecules consisting of a pentose sugar and a nitrogenous base, but no phosphate group, are called nucleosides. In DNA, the pentose sugar is deoxyribose, which differs from the ribose found in RNA by only one oxygen atom.

In both RNA and DNA, the backbone of the molecule consists of alternating sugars and phosphates (—sugar—phosphate—sugar—phosphate—). Bases are attached to the sugars and project from the chain. The nucleotides are joined by covalent bonds in what are called *phosphodiester* linkages between the sugar of one nucleotide and the phosphate of the next ("-diester" refers to the two bonds formed by reacting —OH groups with acidic phosphate groups). The phosphate groups link carbon 3' in one ribose to carbon 5' in the adjacent ribose.

Most RNA molecules are *single-stranded*, consisting of only one polynucleotide chain. DNA, however, is usually *double-stranded*; it has two polynucleotide chains held together by hydrogen bonding between their complementary nitrogenous bases. The two strands of DNA run in opposite directions. You can see what this means by drawing an arrow through the phosphate group from carbon 3' to carbon 5' in the next ribose. If you do this for both strands, the arrows point in opposite directions. This antiparallel orientation is necessary for the strands to fit together.

Uniqueness of a Nucleic Acid Resides in its Base Sequence

Only four nitrogenous bases—and thus only four nucleotides—are found in DNA. The DNA bases and their abbreviations are: adenine (A), cytosine (C), guanine (G), and thymine (T). A key to understanding the structures and functions of nucleic acids is the principle of *complementary base pairing* through hydrogen bond formation. In double-stranded DNA, adenine and thymine always pair (AT), and cytosine and guanine always pair (CG).

Base pairing is complementary because of two factors: the corresponding sites for hydrogen bonding and the molecular sizes of the paired bases. Adenine and guanine are both purines consisting of two fused rings. Thymine and cytosine are both pyrimidines consisting of only one ring. The pairing of a large purine with a small pyrimidine ensures a stable and consistent dimension to the double-stranded molecule of DNA. We'll discuss in more detail how complementary strands of DNA separate and are faithfully copied.

Ribonucleic acids also have four different monomers, but the nucleotides differ from those of DNA. In RNA the nucleotides are termed ribonucleotides. They contain ribose rather than deoxyribose, and instead of the base thymine, RNA uses the base uracil. The other three bases are the same as in DNA.

Although RNA is generally single-stranded, complementary hydrogen bonding between ribonucleotides can take place. These bonds play important roles in determining the shapes of some RNA molecules and in the associations between RNA molecules during protein synthesis. During the DNA-directed synthesis of RNA, complementary base pairing also takes place between ribonucleotides and the bases of DNA. In RNA, guanine and cytosine pair as in DNA, but adenine pairs with uracil. Adenine in an RNA strand can pair either with uracil (in another RNA strand) or with thymine (in a DNA strand).

The three-dimensional appearance of DNA is strikingly regular. Through hydrogen bonding, the two complementary polynucleotide strands pair and twist to form a double helix Compared to the complex and varied tertiary structures of different proteins, this formation seems amazingly regular. But this structural contrast makes sense in terms of the functions of these two classes of macromolecules.

DNA is a purely informational molecule. *The information in DNA is encoded in the sequence of bases carried in its chains.* This sequence is, in a sense, like the tape of a tape recorder. The message must be read easily and reliably, in a specific order. A uniform molecule like DNA can be interpreted by standard molecular machinery, and any cell's machinery can read any molecule of DNA-just as a tape player can play any tape of the right size.

Table 6.1. Distinguishing RNA from DNA

Nucleic acid	*Sugar*	*Bases*
RNA	Ribose	Adenine Cytosine Guanine Uracil
DNA	Deoxyribose	Adenine Cytosine Guanine Thymine

Proteins, on the other hand, have good reason to be so varied. In particular, different enzymes must recognize their own specific "target" molecules, They do this by having a unique three-dimensional form that can match at least a portion of the surface of their target molecules. In other words, structural diversity in the molecules with which enzymes react requires corresponding diversity in the structure of the enzymes themselves. *In DNA the information is in the sequence of bases; in proteins the information is in the shape.*

DNA is a Guide to Evolutionary Relationships

Because DNA carries hereditary information between generations, a series of DNA molecules with changes in base sequences stretches back- through time. Of course, we cannot study all of these DNA molecules, because many of their organisms have become extinct. However, we can study the DNA of living organisms, some of which are judged to belong to lineages that are more ancient than those of other living forms. Comparisons and contrasts of these DNAs add to the evolutionary record.

Closely related living species should have more similar base sequences than species judged by other criteria to be more distantly related. Indeed this is the case. The examination of base sequences confirms many of the evolutionary relationships that have been inferred from the study of microscopic or macroscopic structures or studies of biochemistry and physiology. For example, the closest living relative of humans (*Homo sapiens*) is the chimpanzee (genus *Pan*), which shares more than 98 percent of its DNA base sequence with human DNA.

Confirmation of well-established evolutionary relationships gives credibility to the use of DNA to elucidate relationships when studies of structure are not possible or are not conclusive. For example, DNA studies revealed a close evolutionary relationship between starlings and mocking-birds that was not expected on the basis of studies of anatomy or behaviour. DNA studies support the division of the prokaryotes into two kingdoms (Eubacteria and Archaebacteria). Each of these two groups of prokaryotes is as distinct from the other as either is from the other four kingdoms into which living things are classified. In addition, DNA comparisons support the hypothesis that certain subcellular compartments of eukaryotes (the organelles called *mitochondria* and *chloroplasts*) evolved from early bacteria that established a stable and mutually beneficial way of life inside larger cells.

Interactions of Macromolecules

We have been treating the classes of macromolecules as if each were completely separate from the others. In cells, however, certain macromolecules of different classes may be covalently bonded to one another. Proteins with attached oligosaccharides are called *glycoproteins* (*glyco-*, "sugar"). The specific oligosaccharide chain determines the placement of a glycoprotein within the cell.

Other carbohydrate chains bind to lipids, resulting in *glycolipids*, which reside in the cell surface membrane, with the carbohydrate chain extending out into the cell's environment. In humans, the substances that determine the ABO blood types are carbohydrates attached to either proteins or lipids. When a human body cell becomes cancerous, the glycolipids and glycoproteins on the cell surface are modified, and these changes serve as a recognition signal for the body defenses to destroy the abnormal cell.

Not all the associations among macromolecules or between a macromolecule and a smaller molecule are covalent. Weaker linkages such as ionic bonds, hydrogen bonds, and even van der Waals forces may also be involved. Enzymes bind to their reactants using a variety of weak bonds. Ionic bonds link proteins to DNA, forming *nucleoproteins* that regulate the activities of DNA. Still other proteins, in combination with cholesterol and other lipids, form *lipoproteins*. The lipoproteins make it possible to move very hydrophobic lipids through water-rich environments such as the blood and tissue fluid and deliver cholesterol to all the body's cells.

7

MOLECULAR ENGINEERING

A *molecule* is a collection of (one or more) atoms which are bound together by their mutual interactions for long enough to be observed as an entity. There is an enormous range in molecular stabi!ities and lifetimes, with some molecules existing only for the duration of an experiment lasting 10^{-12} [a trillionth of a second] or less, while others may remain intact for billions of years molecules are sometimes classified by the number of atoms they contain (E.g., monatomic, diatomic, triatomic); those with three or more atoms are generally termed polyatomic. Molecules range in size from monatomic species (as are found in gaseous helium or argon) to macroscopic aggregates such as single crystals of diamond or quartz, and polymers), an din complexity from simple atoms to proteins, enzymes, and nucleic acids. Every known kind of atom is found in at least one diatomic (or larger) molecule; some atoms (e.g., carbon and hydrogen) are found as constituents of millions of different molecules. Molecules that contain carbon (with a few exceptions) are called organic; all other are inorganic, but these terms no longer imply a connection with living organisms. It is generally recognized that *engineering* is "the art or science of making practical application of the knowledge of pure science."

NANOTECHNOLOGY IS MOLECULAR ENGINEERING

Nanotechnology is the art and science of building complex, practical devices with atomic precision. In an effort to construct useful machines, nanotechnologists apply the techniques of engineering to the knowledge generated by the sciences that study molecular structures. If we focus or attention on the developing capacity of these sciences in the closing years of the twentieth century, we can observe the birth of nanotechnology. In the next century, molecular engineering will emerge as a multitrillion dollar industry that will dominate the economic and ecological fabric of our lives.

Molecular engineering begins with an overview of the scale of molecular objects. Then, following a brief introduction to atoms and molecules, and a cursory review of Nanotechnology as an evolving discipline. It highlights recent research efforts that are leading the dream of molecular engineering into reality.

SCALE

To fully understand molecular engineering—to be a nanotechnologist —requires study in several fields, including physics, chemistry, molecular biology, and computer science. But this book is not a technical exposition of nanotechnology. Rather it is an invitation to well-founded social dialogue as we collectively imagine and choose among different evolutionary paths. As such, the material presented does not focus on the analysis and mathematics required presented does not focus on the analysis and mathematics required to actually design and build molecular machines. But we cannot escape the fact that it is precisely the scale at which nanotechnology operates that gives it its power. So, unless we intend to collapse into the simple repetition of superlatives, we need to come to terms with the dimensions of molecular objects.

The goal of nanotechnology is the construction of a wide range of artifacts whose components are reasonably measured in nanometers, or *billionths* of a meter. At first it may seem impossible to imagine to billionth of a meter, or an apparatus with trillions of parts executing movements in quadrillionths of a second. But the human mind is a truly marvelous thing. With it we can imaging not only a frabjous day and a slain Jabberwock but also machines with precision-crafted molecular components. And while the former seem destined to remain figments, the latter images may someday be as real as you or me.

An imaginative exploration of nanotechnology requires the two essential dimensions of any material system—space and time—as they relate to the molecular world.

First, how big are atoms? If, as the Greek philosopher Protagoras argued, Man is the measure of all things," it is not surprising that we use a system of measurement with a unit of length roughly equal to the size of a human being. The metric system, which was proposed by

Gabriel Mouton, Vicar of Lyons, in 1670, and adopted in France in 1795 subsequent to the French Revolution, is founded on two fundamental units: the ***meter*** and the ***kilogram***. The ***meter*** is about forty inches long and a kilogram weighs a little over two pounds. Because the ***metric system*** is a decimal system, all units can be expanded by multiples of ten. This allows for the use of a collection of prefixes, each of which can be affixed to a given unit, that indicate every third order of magnitude:

exa-	10^{18}	1,000,000,000,000,000,000	quintillion
peta-	10^{15}	1,000,000,000,000,000	quadrillion
tera-	10^{12}	1,000,000,000,000	trillion
giga-	10^{9}	1,000,000,000	billion
mega-	10^{6}	1,000	thousand
	10^{0}	1	
milli-	10^{-3}	1/1,000	thousandth
micro-	10^{-6}	1/1,000,000	millionth
nano-	$\mathbf{10^{-9}}$	**1/1,000,000,000**	**billionth**
pico-	10^{-12}	1/1,000,000,000,000	trillionth
femto-	10^{-15}	1/1,000,000,000,000,000	quadrillionth
atto-	10^{-16}	1/1,000,000,000,000,000,000	quintillionth

The diameter of a single atom is a bit larger than one-tenth of a *nanometer* (nm). Consequently, modest atomic constructions are most reasonably measured with these units. The DNA molecule, for example, is about 2.3 nanometers wide. Large, complex molecular structures may be measured in *micrometers*, or *microns* (μm).

Although most dimensions in this book are given in *microns* and *nanometers*, other scales are common in physics and biochemistry: the *angstrom*, the *dalton*, and the *atomic mass unit*. The *angstrom* (Å) named after the Swedish physicist Anders Jonas Angstrom (1814–1874), is one tenth of a nanometer, of 10^{-10} meter. The *dalton*, named after the English chemist John Dalton (1766–1844), is a unit of mass nearly equivalent to that of a single hydrogen atom. The *kilodalton* (kd)—equal to a thousand daltons—is often used to indicate the molecular weight of moderate-to-large biomolecules. *Rhodopsin*, for example, is a medium-sized structural protein that holds in place a small, 24-atom molecule called *retinal*. Retinal is responsible for the primary visual event in the rod cells of the retina. Rhodopsin weights about 40,000 hydrogen atoms. The *atomic mass unit* (amu) is another run it that indicates the approximate mass of one hydrogen atom.

To get an idea of how small atoms are, Kenneth Ford offers the following image. "To arrive at the number of atoms in a cubic centimeter of water (a few drips), first cover the earth with airports, one against the other. Then go up a mile or so and build another solid layer of airports. Do this 100 million times. The last layer will have reached out to the sun and will contain some 10^{16} airports (ten million billion). The number of atoms in a few drops of water will be the number of airports filling up this substantial part of the solar system. If the airport construction rate were *one million* each second, the job could just have been finished in the known life time of the universe (sometime over ten billion years)."

While atoms are incredibly small we have, in the past decade, built the tools to not only "see" individual atoms but to manipulate them as well.

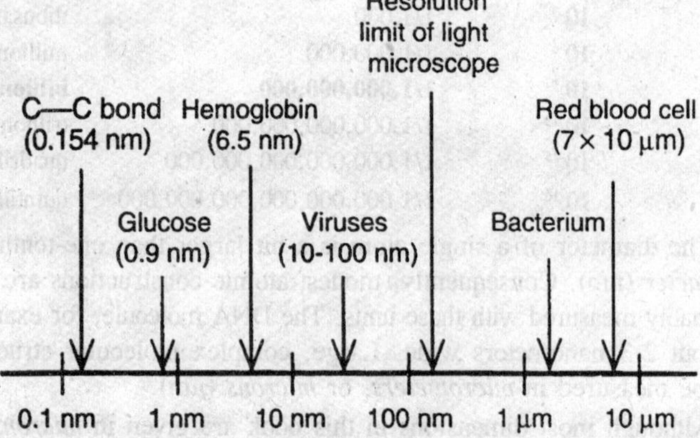

Fig. 7.1. The relative size of atoms, biomolecules, bacteria, and cells.

The second dimension we need to imagine is the *speed* with which atoms move. In the atmosphere, oxygen, nitrogen, carbon dioxide, and other simple molecules zip around at up to ten times the speed of sound, or about 7,000 miles an hour. But the local transformations that molecules are capable of are much more important than their gross physical movement as gaseous particles. For example, the primary event in vision—the mechanical transformation of the molecule retinal from bent to straight—occurs within 200 *femtoseconds* (*quadrillionths* of a second) after begin struck by a photon.

In watery solutions, such as those found in the interior of most plants and animals, the general jostling caused by thermal excitation bumps protein-sized molecules into each other from all angles very

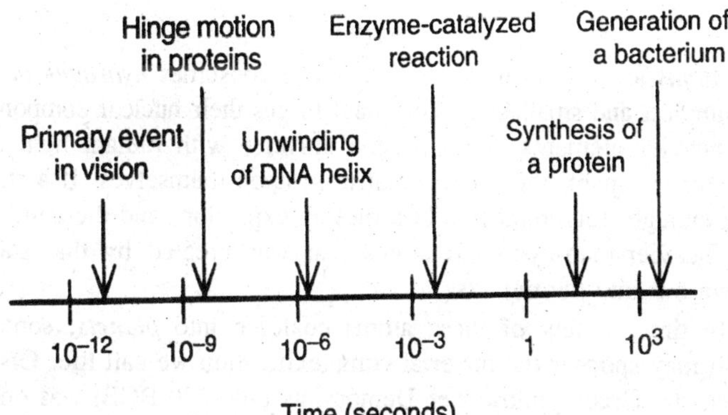

Fig. 7.2. The relative rates of some biological processes.

rapidly. This bumbling, stumbling dance allows molecules to explore all possible "*mating*" configurations with the other molecules in their local environment. By variously constraining and controlling the chaos of such wild interactions, biological system generate the event we call life. And it is exactly this mechanism of molecular *self-assembly* that may lead to the construction of complex structures designed by human engineers.

One final not on scale. Although an atom is small, the nucleus is much much smaller, and of relatively little interest to nanotechnology. Kenneth Ford again: "To picture the nucleus, whose size is about 10^{-4} to 10^{-5} of the size of an atom, one may the atom expanded to say, 10,000 feet (10^4 feet)) or nearly two miles. This is about the length of a runway at a large air terminal such as New York International Airport. A fraction 10^{-4} of this is one foot, or about the diameter of a basketball. A fraction 10^{-5} is ten times smaller, or about the diameter of a golf ball. A golf ball in the middle of the New York International Airport is about as lonely as the *proton* at the center of a hydrogen atom. A basketball would correspond to a heavy nucleus such as uranium."

The *electron* cloud that surrounds the nucleus determines the effective size of an atom, not the *protons* and *neutrons* at its core. Nanotechnology does not smash atoms to extract the energy held in the nucleus, as happens in nuclear power plants and atomic weapons. Nanotechnology is constructive; it snaps atoms together like Lego building blocks to build molecular structures in processes that are similar to, but potentially much more flexible and powerful than, the processes used by biological systems.

Atoms

Atoms are forged in stars. Stellar fire consumes *hydrogen atoms*, the simplest and smallest element, and forges their nuclear components into heavier elements. Stars fill the universe with radiant light until their fuel is spent, then they collapse in upon themselves. If a star is large enough, this implosion becomes an explosion, and the collection of a heavier-than-hydrogen atoms that was created by that star is thrown out into the universe.

In time, a few of these atoms coalesce into *planets*, some of which may sponsor the material complexification we call life. On this planet, the Greek philosopher Democritus (460-370 BCE) was one of the first living creatures to suggest that the world consisted of very small but indivisible and indestructible pieces. These pieces were called "*atoms*" (*atoms*, indivisible: *a*-, not, + *temnein*, to cut). Democritus felt that everything that we see could be adequately explained as an aggregate or combination of atoms. And while "there are many different kinds of atoms, that is to say, they are not all the same size and shape..., no atom is large enough to be seen; conversely, no atom is so small as to have no dimensions at all." Although we have moved from a philosophical to a scientific understanding of atoms, it is startling to realize that this ancient intuition of the underlying structure of matter is almost sufficient to comprehend the consequences, if not the precise machinations, of molecular engineering.

Human understanding of elemental structures evolved slowly until the late eighteenth and early nineteenth century when Antoine Lavoisier (1743–1794) proposed that elements retained their weight regardless of the compounds that they formed, and John Dalton (1766–1844) connected this concept with an *atomic theory*. Dalton is generally given credit for establishing the modern atomic theory with the publication of *A New System of Chemical Philosophy* in 1808, coincidentally the same year Ludwig van Beethoven first performed his Fifth Symphony. Dalton argued that the number of atoms in a gas, liquid, or solid, while numerous, must be finite, and that every atom of a particular element must be identical in nature. "His atoms were no longer smallest particles with some general and rather vague physical properties, but atoms endowed with the properties of chemical elements."

Dalton wrote, "Whether the ultimate particles of a body of water, are all alike, that is of the same figure, weight, at cetera, is a question of some importance. From what is known, we have no reason to apprehend a diversity in these particulars: if it does exist in water it must equally exist in the elements constituting water, namely, hydrogen and oxygen. Now it is scarcely possible to conceive how the aggregates of dissimilar particles should be so uniformly the same. If some of the particles of water were heavier than others, if a parcel of a liquid on any occasion were constituted principally of these heavier particles, it must be supposed to effect the specific gravity of the mass, a circumstance not known. Similar observations may be made on other substances. Therefore we may conclude that *the ultimate particles of all homogeneous bodies are perfectly alike in weight, figure, et cetera.* In other words, every particle of water is like every other particle of water; every particle of hydrogen is like every other particle, at cetera."

While Dalton's assumption that the simplest numerical combinations were in fact the actual ratios with which atoms combined led to some erroneous conclusions (e.g., that the smallest particle of water consisted of one oxygen and one hydrogen atom, rather than two hydrogens), his "new system" satisfied all known results and opened the door to rapid development of a modern atomic theory.

In passing, it is worth noting that the drawings Dalton used to present his theory in 1808 are flawed in that compounds are shown to consists of a planar aggregates of circular atoms rather than three-dimensional collections of spheres. While atoms are clearly more complex than simple balls, the best first approximation of molecular form is that of closely packed spheres in three-space.

Sixty years after Dalton's landmark publication, a total of sixty-three different elements had been effectively isolated and described in the emerging technical literature. The time was ripe for the imposition of order on this growing collection. Certain regularities were beginning to emerge and, in 1866, John Newlands attempted to articulate an overreaching pattern. He earned himself the contempt and ridicule of the English Chemical Society for proposing that the elements could be meaningfully arranged by *molecular weight* in groups of eight—as notes in a scale—with every eighth element similar in many respects to the others so grouped. Newlands was right that there was a repetitive theme in the elements, but his structure was not sufficiently all encompassing, and his analogy to music was far too strange for his audience to appreciate. Newlands's career all but ended.

Three years later, as the first Atlantic-Pacific railway line was completed at Promontory Point, Utah, the Russian chemist Dmitri Mendeleev not only announced a cyclic pattern for organizing all known elements but he also accompanied his announcement with the declaration

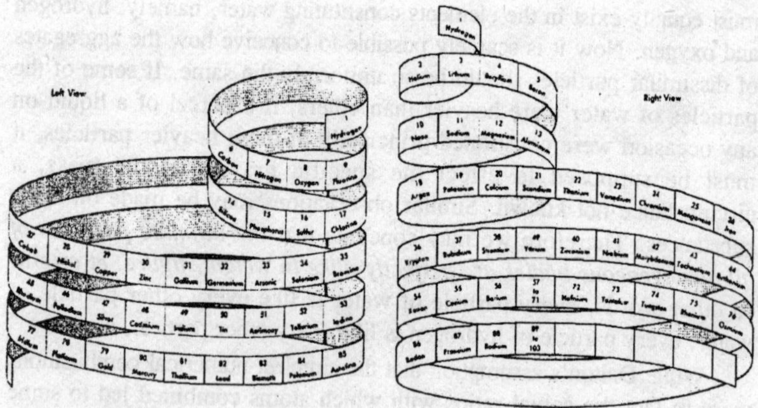

Fig. 7.3. The periodic table. "The chemical elements have been arranged in sequence on a ribbon and coiled in a helix.

that three specific elements had yet to be discovered. As he constructed it, the "*periodic*" table contained several gaps, yet he was bold enough to predict not only the atomic weight but also the physical and chemical properties of several of the elements that would eventually fill out the table. As these elements were discovered, Mendeleev's place in history was secured. The first missing element—*gallium*—was discovered four years later, in 1875, just as James Clerk Maxwell opened the next level of exploration by noting that atoms themselves were rather more complex than had been imagined.

"Having thus indicated a new mystery of Nature," wrote Mendeleev, "which does not yet yield to rational conception, the periodic law, together with the revelations of spectrum analysis, have contributed to again revive an old but remarkably long-lived hope—that of discovering, if not by experiment, at least by a mental effort, the *primary matter*—which had its genesis in the minds of the Grecian philosophers, and has been transmitted, together with many other ideas of the classic period, to the heirs of their civilization."

Molecules

Molecules are collections of atoms bound to one another. They compromise almost everything that we interact with. We breathe them, eat them, wash with them, and even think and feel with them. Solid matter contains on the order of 10^{23} molecules per cubic centimeter. Chemists estimate that there are 12 million specific molecular compounds on record, to which some 500,000 new compounds are added each year. This section introduces the barest few of this vast horde.

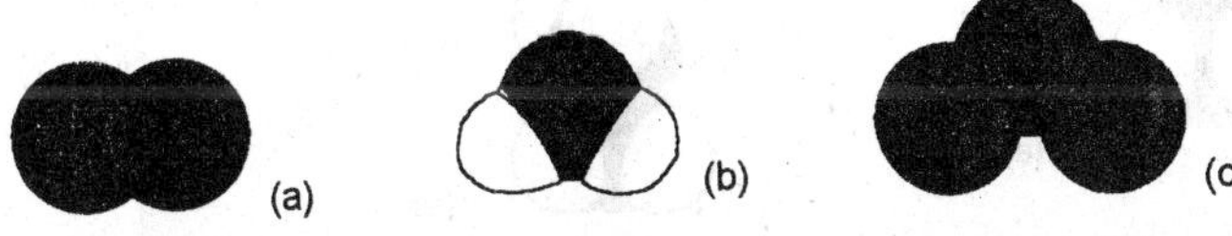

Fig. 7.4. (a) Oxygen (O_2), (b) water (H_2O), and (c) ozone (O_3).

Atmospheric oxygen is diatomic. Water, H_2O, is triatomic, consisting of two atoms of hydrogen and one of oxygen. Ozone is also triatomic, consisting of three oxygen atoms. When ultraviolet radiation from the sun strikes O_2 molecules (and other oxygen-containing molecules) in the upper atmosphere, it blasts the oxygen atoms apart. Single oxygen atoms then bond to an existing oxygen molecule (O_2) to form ozone. Ozone molecules are also torn apart by ultraviolet radiation. Each of these processes absorbs particular wavelengths in the ultraviolet portion of the electromagnetic spectrum. Bombarded by solar radiation, this frenzied activity proceeds continuously in a 10-kilometer layer between 25 and 35 kilometers above the Earth's surface. "The absorption of ultraviolet radiation by the gas is so efficient that at wavelengths near 250 nanometers, in the ultraviolet, only one part in 10^{30} of the incident solar radiation penetrates the ozone layer."

Deletion of the ozone layer is dangerous for living things because the molecule that contains the instructions for building our bodies, DNA, responds dramatically to ultraviolet radiation. The DNA molecule has a natural harmonic at the frequency of certain ultraviolet light. When excited by this radiation, DNA vibrates excessively and falls apart like a bridge in an earthquake. Even though the DNA is healthy cells is constantly under repair, cancers such as melanomas can result when the disruption is too severe. In most cases, cancer is simply the result of a cells' best effort to follow a fouled set of instructions.

Methane is a simple, flammable hydrocarbon made from five atoms. Ethane and octane, more complex hydrocarbons, consist of eight and twenty-six atoms, respectively. "The octane molecule results when we continue the process that leads from methane to ethane. Now a sufficient number of —CH_2—units have been introduced into the original C–H bond of methane to make the chain eight carbon atoms long (hence octane). Hydrocarbon molecules that contain half a dozen or so carbon atoms interact just strongly enough with each other to given a liquid at room temperature, so they are convenient to transport in tanks. But the liquid is still volatile and not to viscous to form a spray in the carburetor of an engine. Octane is representative of the size of the hydrocarbon molecules present in a gallon of gasoline....

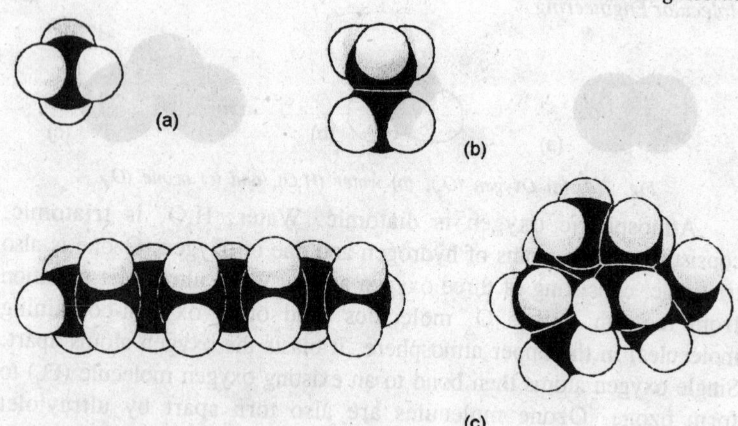

Fig. 7.5. (a) Methane (CH_4), (b) ethane (C_2H_6), and (c) octane (C_8H_{18}) in two configurations.

Diesel fuel is less volatile: Its molecules are typically hydrocarbons with about sixteen carbon atoms in a chain... Every hydrocarbon chain is actually a zigzag of carbon atoms and is flexible as well; moreover, each atom can be twisted around the bond joining it to its neighbour. A gallon of gasoline therefore contains some octane molecules that are rolled up into a light ball, others are stretched out but still zigzag and others in the various intermediate configurations. Octane molecules are constantly writhing and twisting, rolling and unrolling, so that a gallon of gasoline is more like a can of molecular maggots than a box of short sticks."

The double helix of DNA, which consists of thousands of adenine-thymine and guanine-cytosine pairs stacked one atop the other, is 2.3 nm wide. Human DNA, which is found in almost every cell in the

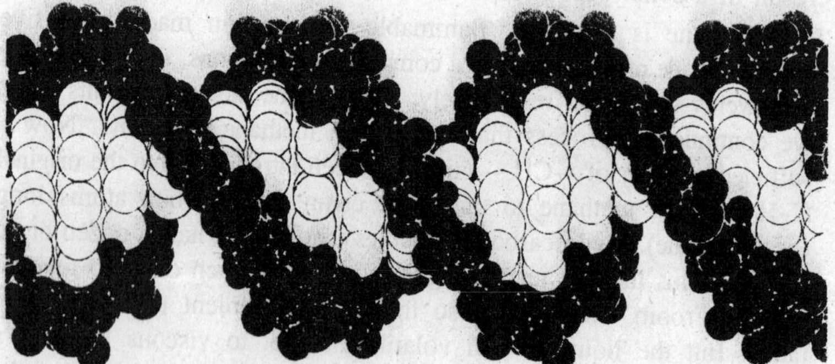

Fig. 7.6. Eighteen base pairs of DNA, approximately 6 nm long and 2.3 nm wide.

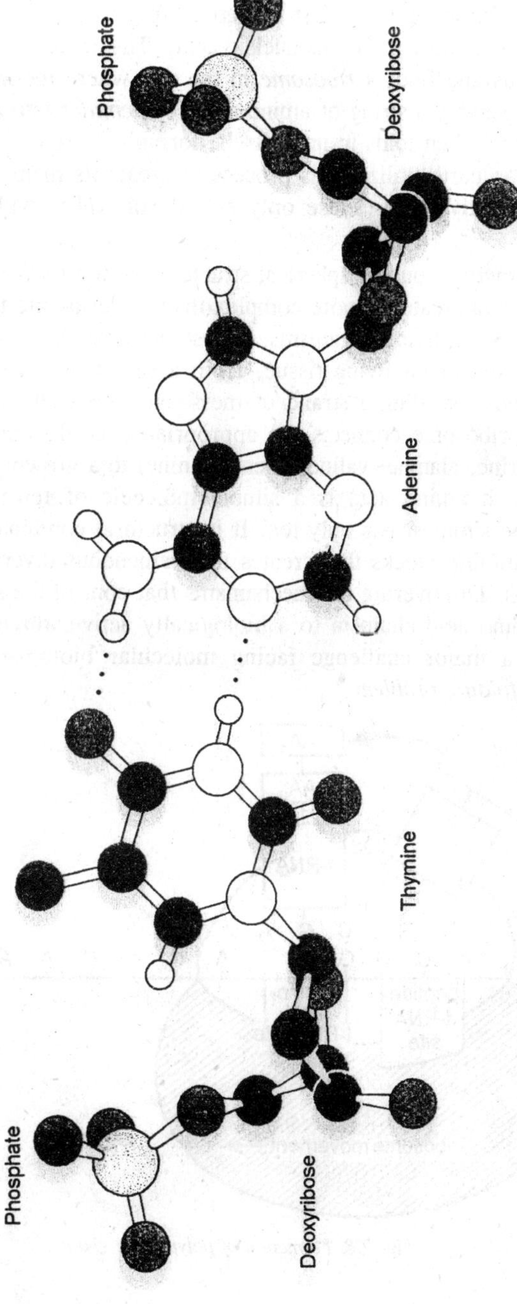

Fig. 7.7. A single base pair of deoxyribonucleic acid or DNA. Adenine (26 atoms) is hydrogen bonded (dotted lines) to thymine (23 atoms) in this base pair. The other possible base pair consists of guanine and cytosine molecules.

body, is nearly 10 meters long. To make the molecular components of the body, information is first copied from a strand of DNA onto a strand of *messenger RNA* (ribonucleic acid). This molecular chain leaves the nucleus and finds a *ribosome* in the cell where the information is used to assemble a string of amino acids of *peptides* into a *polypeptide* chain, which then folds in upon itself, forming a protein. Every living creature on earth utilizes this process to create its many components. (Some primitive viruses use only strands of RNA, without a DNA source.)

Ribosomes—roughly spherical structures with a mass of about 4200 *kilodaltons* in creatures more complex than bacteria—are the molecular factories for all living organisms. Ribosomes assemble proteins, which make up almost all living tissue, from a set of just twenty different amino acids. Reading a strand of messenger RNA, three "letters" at a time a ribosome connects the appropriate free-floating amino acid (E.g., serine, alanine, valine, phenylalanine) to a growing polypeptide chain. Each amino acid is a simple molecule of fewer than thirty atoms; the simplest has only ten. It is structural combination of these simple building blocks that creates the tremendous diversity of living organisms. Discovering the mechanisms that control the self-assembly of an amino acid chain in to a biologically active, three-dimensional from is a major challenge facing molecular biologists today: the "*protein-folding problem.*"

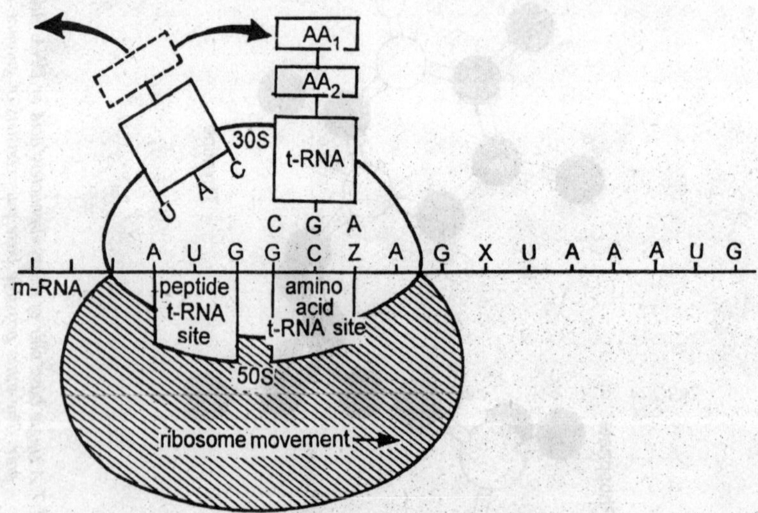

Fig. 7.8. Formation of polypeptide chain.

Ribosomes are particularly interesting because they demonstrate that a simple molecular machine can, given the appropriate instruction, assemble a vast array of other molecular structures. In fact, because ribosomes are themselves protein complexes, each ribosome is assembled by other ribosomes. From the perspective of nanotechnology, most of molecular biology, including the study of ribosomes, can be seen as a massive reverse-engineering project, where the machines in question are utterly without documentation.

While ribosomes and other naturally evolved mechanism of protein formation are today exploited by recombinant DNA technology and genetic engineering, the goal of molecular nanotechnology is the construction of *general-purpose assemblers* able to build molecular structures with any conceivable configuration—given the necessary instructions and assuming the structure is chemically stable. Assemblers with the capacity to build assemblers that are identical to themselves are called *replicators*.

Another biological system of particular interest to nanotechnology is the molecular reaction used to power most changes in living cells. The partial decomposition of adenosine triphosphate (ATP) to adenosine diphosphate (ADP) releases a very small amount of energy or heat. This energy is used to drive other chemical reactions that would not otherwise occur. For example, ribosomes consume a great many ATP molecules (turning them into ADP) as they coordinate the bonding reactions that assemble polypeptide chains. "The gearing of the reactions is such that the polypeptide is built at the expense of greater disorder elsewhere. To sustain these reactions, we need to eat, for we must rebuild the ATP from the ADP, using the energy of reaction of reactions stemming from respiration [for oxygen] and digestion [for sugars]. In a green leaf—one of the sources of the food that fuels us—the rebuilding of ATP from ADP is achieved by photosynthesis, and there the distant waterfall that drives the reaction is nuclear fusion in the sun. All life is the outgrowth of this intricately geared collapse into cosmic chaos."

It is worth noting that humans have discovered and learned to use a molecule that is sufficiently similar to ATP that it can mimic its effects. Billions of us consume coffee, tea, and cola drinks because they temporarily increase our allotment of biological energy by providing our cells with caffeine molecules. Cocaine molecules on the other hand, which used to be added to Coca Cola and other drinks, act like amphetamine. Rather than augmenting the supply of cellular ATP, amphetamine-like molecules insert themselves directly between neurons, mimicking the neurotransmitter norepinephrine.

On examination, all so-called *drugs* intervene in the processes of our bodies at the molecular level. A few simple molecules, such as nicotine, morphine, aspirin, cocaine, tetrahydrocannabinol (the active component of cannabis), and ethyl alcohol are responsible for the movement of billions of dollars a year—and none contain more than sixty atoms; most have fewer than thirty. While nanotechnology speaks to the possibility of a radically new level of molecular control, it is clear that we already live in a world of powerful molecules.

Genealogy of Nanotechnology

The companion fields of nanotechnology and artificial-life studies can be usefully thought of as being the *inside-out* of each other. Each is dependent on the implicates the other; each is essentially useless and meaningless without the other. Nanotechnology addresses a collection of problems from a mechanical perspective, while artificial life—understood as the goal of an evolving effort in computer science to construct intelligent information systems—addresses essentially the same problems from an informational, systems-theoretic perspective. Nanotechnology without artificial life can be little more than precise chemistry; artificial life without a nanotechnological embodiment offers little more than barren simulations.

Considering linguistic roots for a moment, we can see that "technology" derives from two Greek words, *tekhne*, meaning skill or art, and *logos*, meaning word or speech. "Nanotechnology" therefore indicates the discourse, hence the science, theory, or study of the skills required to craft matter at the nanometer scale. "Artificial" derives from the term "artifice," and comes to us from the Latin *artifex*, meaning craftsman: *ars* (stem *art-*), art + *-fex*, maker. "Life" is a much younger word, appearing in Old English as *lif* and seems to indicate the idea, "that which continues."

A thorough genealogical treatment of nanotechnology and artificial life would therefore require tracing the evolution and development of not only the specific fields of chemistry, biology, engineering, and mathematics—especially applied mathematics as embodied in computer science—but it would also require an exploration of the anthropological roots of *homo faber*; the human artist, craftsman, and storyteller. Clearly, a brief overview must suffice.

If we allow that the primordial nanoscale "*machine*" is the double helix itself, we can begin by recalling a provocatively titled volume that appeared within months of our entry into the nuclear age. In 1945, on July 16, the United States detonated the first atomic weapon near Alamogordo, New Mexico. In less than a month, atomic weapons were exploded over Hiroshima and Nagasaki. While the first bomb was in development, Erwin Schrodinger, a physicist who received the Nobel Prize in 1933 for discovering new forms of atomic energy, published a slim volume that asked a simple question: *What is Life?*

Schrodinger's concern was not primarily philosophical. The question he asked was, "How can the events in space and time which take place within the spatial boundary of a living organism be accounted for by physics and chemistry? Schrodinger's essay is essentially an inquiry into a materiality of cellular life. "Adopting a conjecture already current, that a gene, or perhaps even an entire chromosome fiber, is a giant molecule, he takes the further stop of supposing that its structure is that of an *aperiodic* solid or crystal... Like other crystals the chromosome can reproduce itself. But it also has another important attribute which is unique: its own complex structure (the energy state and configuration of its constituent atoms) forms a 'code-script' that determines the entire future development of the organism." Schrodinger thus anticipated a key characteristic of the molecule of heredity—deoxyribonucleic acid or DNA: its capacity to act as a set of instructions for the material construction of living forms.

In the same year, 1945, John von Neumann published his *First Draft of a Report on the EDVAC*, which described the basic concepts of the modern computer. Given the apparent similarities between organic systems and logical computation, it was inevitable that the mathematicians responsible for designing the first electronic computers would investigate the information-processing capabilities of both living creatures and engineered *automata*.

Von Neumann was an extraordinarily brilliant mathematician. In addition to consulting on the Manhattan Project and pioneering what would become the standard architecture for computing machinery, he occupied himself with what he called *automata theory*, formulating axioms and proving theorems about assemblies of simple elements that might in an idealized way represent either possible circuits in man-made automata or patterns in organisms. Although he was stimulated by the problems of computers and organisms and continued to take account of them, he found the new abstract mathematics intrinsically interesting. One question with which he dealt in his automata theory is, if one has a collection of connected elements computing and transmitting information, i.e., an automaton, and each element is subject to malfunction, how can one arrange and organize the elements so that the overall output of the automation is error free? The other problem was closer to genetics,... that of constructing of formal models of automata capable of reproducing themselves. These would be models of a basic element of 'life'.

In 1953, as Tenzing Norgay and Edmund Hillary reached the summit of Mount Everest, James Watson and Francis Crick developed a convincing model for the giant molecule of heredity," thereby validating Schrodinger's intuition. This new understanding sparked an explosion of investigations into the molecular mechanisms of organic life.

As biochemists busily studied the mechanisms of life as we know it, the physicist Richard Feynman pondered the physical limits of machinery as we might build it. In 1959, Feynman entertained an audience at the California Institute of Technology with a visionary presentation entitled, "There's plenty of Room at the Bottom." In this talk, Feynman described a recursive process whereby very small machines could be built: each generation of machine tools would craft another regeneration with finer capabilities. At the conclusion of his talk, he predicted the arrival of atomically precise machinery. "I am not afraid to consider the final question as to whether, ultimately—in the very great future—we can arrange atoms the way we want; the very atoms, all the way down!

In the years that followed, chemists and biologists focused on untangling the molecular structures that constitute materiality from the "*bottom up*," while physicists and electrical engineer is devoted their efforts to building ever smaller machines from the "*top down.*" Both efforts resulted in tremendously powerful technologies. Molecular science produced revolutionary medicines as well as the novel materials that provide the texture of so much of the modern world—including nylon, Tyvek, Teflon, and super glue—while micromachinists, after creating the first transistor in 1948, learned to build logic and computation machines with micron-scale components, generating thereby a global industry second only to agriculture.

The recent confluence of these two monumental efforts has produced an epochal *cross-fertilization* of knowledge—and the inevitable conceptual turbulence of two colliding world views. While top-down engineers such as computer chip designers build a fairly small number of complex machines as minutely as possible, chemists and other bottom-up technologists build relatively simple but atomically precise machines by the billions. Nanotechnology rises out of this confluence and aims at building complex, atomically precise machines by the trillions.

One of the most important developments for both approaches to the design and control of submicron systems was the invention of the *scanning tunneling microscope* (STM) in 1981. This device, first developed by Gerd Binnig and Heinrich Rohrer at IBM's Zurich Research Labs, gave us the first direct images of individual atoms. "The technique used... involves an ultrafine stylus that hovers slightly above a conducting surface and senses topographic details via tiny fluctuations in the 'tunneling current' that forms between the stylus and the surface. In effect, the STM senses the outer surface of the electron cloud that defines at atom. The STM, which earned Binning and Rohrer the 1986 Nobel Prize in physics, can imagine only conducting surfaces, but this limitation was overcome with the development of the *atomic force microscope* (AFM), in 1986, which can image nonconducting surfaces with similar resolution.

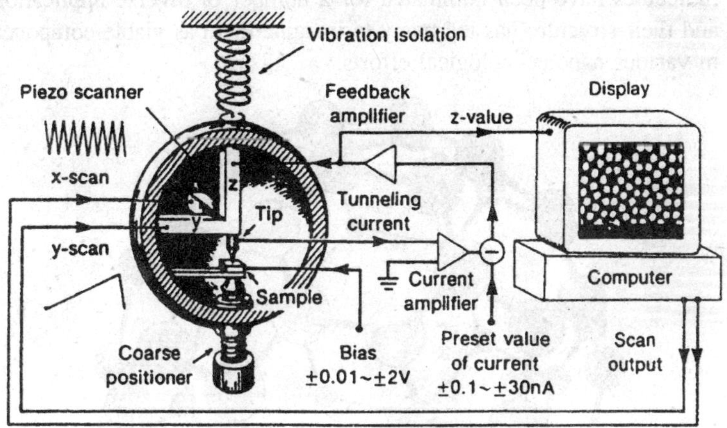

Fig. 7.9. Scanning tunneling microscope.

Also in 1981, K. Eric Drexler, then a researcher at the Space Systems Laboratory of MIT, published a paper entitled, "*Molecular engineering*: An approach to the development of general capabilities for molecular manipulation," in which he argued that the natural mechanisms of protein synthesis demonstrate the feasibility of human-engineered molecular machines. "To deny the feasibility of advanced molecular machinery, one must apparently maintain either (i) that design of proteins will remain infeasible indefinitely, or (ii) that complex machines cannot be made of proteins, or (iii) that protein machines cannot build second-generation machines. Drexler argued that none of these objections can be sustained. In 1982, Drexler introduced the concept of molecular engineering to a popular audience with the publication of "When molecules will do the work" in *Smithsonian* magazine.

In 1985, the discovery of a new form of carbon molecule stunned the scientific community. Carbon participates in a huge variety of molecules, but only two-carbon forms were previously known: graphite, which consists of two-dimensional sheets, and diamond, which is a three-dimensional network of interlinked atoms. In contrast, the molecule *buckminsterfullerene* contains exactly sixty carbon atoms in the shape of a soccer ball. In 1991, *buckyballs*, as they came to be known, were heralded as the "*Molecule of the Year*" by the American Association for the Advancement of Science and appeared on the cover of *Scientific* American. Named after R. Buckminster Fuller, these roughly spherical molecules have been nominated for a number of diverse applications, and their structure has led many to imagine them as viable components in various nanotechnological efforts.

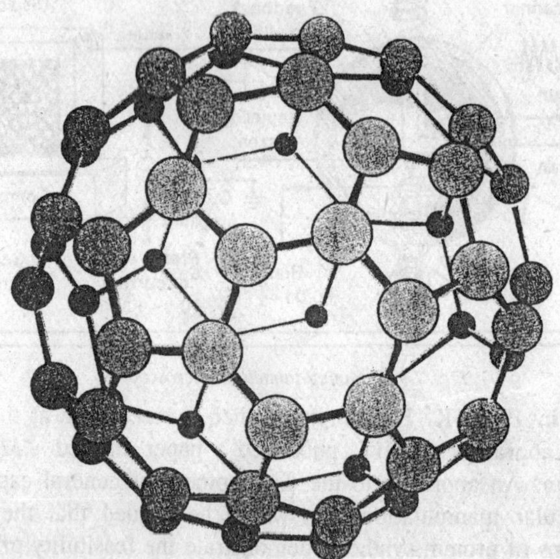

Fig. 7.10. Buckminsterfullerence (C_{60}). A third form of pure carbon.

In 1986, while Binning and Rohrer received their award in Stockholm for revealing to the world images of individual atoms, K. Eric Drexler published the first full-length book on the potential of molecular engineering for a popular science audience. The *Engines of Creation: Challenges and Choices of the Last Technological Revolution* introduced most of the writers in this volume, as well as many others, to the field of nanotechnology.

In the following year, 1987, the first workshop on artificial life was held in Los Alamos, New Mexico. Jointly sponsored by the Center for Nonlinear Studies at the Los Alamos National Laboratory, the Santa Fe Institute, and Apple Computer, Inc., "the workshop brought together 160 computer scientists, biologists, physicists, anthropologists, and other assorted '-ists,' all of whom shared a common interest in the simulation and synthesis of living systems." Several papers presented at this ground-breaking conference demonstrate the confluence of top-down and bottom-up approaches to molecular systems design, including a survey article by Conrad Schneiker, "Nano-Technology with Feynman Machines: Scanning Tunneling Engineering and Artificial Life." Schneiker suggests that "there are many ways that nanotechnology can eventually be applied to the development of artificial life... (1) we can start with a completely natural life form and gradually transform it (bootstrap it) into a totally artificial life form by using molecule-by-molecule replacement. (2) We can develop a hybrid living system that incorporates some nanotechnology for computing functions, and some microtechnology for artificial replication."

In 1988, Richard Feldmann, a computer scientist at the National Institute of Health (NIH), presented a paper, "Applying Engineering Principles to the Design of a cellular Biology," in which he argued that building a biology is a reasonable goal for the scientific community. Feldmann voices an optimism familiar to computer scientists: "With exponentially increasing computer power, it will take far less than 80 years to be able to design and implement a biological system. The issue seems to be simply one of deciding we want to do a project like this, not the technological complexity of the project *per se*.

In the same year, Hans Moravec, the Director of the Mobile Robot Laboratory at Carnegie Mellon University, published *Mind Children: The future of Robot and Human Intelligence*. In this book he describes "A nanocomputer [that] might have a processing speed of a trillion operations per second. With millions of processors crammed onto a thumbnail-size chip, my human-equivalence criterion [for intelligence] would be bested more than a millionfold!"

Two years after the first conference on artificial life, the first international conference on nanotechnology was held in Palo Alto,

California, sponsored by the Foresight Institute (founded by Drexler) and the Global Business Network, and hosted by the Department of Computer Science at Stanford University. The volume that resulted from that conference, which I had the pleasure of editing, presents a broad range of technologies that contribute to nanotechnology as well as several perspectives on the consequences of success. Here," as one reviewer wrote, "in a highly accessible format, is a discussion of atomic probe microscopes, self-assembly in molecular crystals, protein engineering, micromachining and much else.

In July of 1990, the English Institute of Physics launched a new journal called *Nanotechnology*. The ground-breaking issue included articles on various submicron technologies, including, "The Scanning Tunneling Microscope as a Tool for Nanofabrication." Also in 1990, IBM set a record for miniaturized publicity—and brought nanotechnology to the attention of the popular press—by spelling out their company logo with thirty-five xenon atoms on a nickel crystal.

In 1991, several companies announced their intentions to invest in nano-scale research. IMB's president for Science and Technology, J.A. Armstrong, wrote, "I believe that nanoscience and nanotechnology will be central to the next epoch of the information age, and will be as revolutionary as science and technology at the micron scale have been since the early '70s... Not only do we have the ability to make such nanostructures, but, as an outgrowth of the invention of scanning tunneling microscopy, we have the micromechanical ability to manipulate, as well as to see and measure, these structures... Indeed, we will have the ability to make electronic and mechanical devices atom-by-atom when that is appropriate to the job at hand. IN Japan, the Ministry of International Trade and Industry (MITI) also announced the funding of a broad nanotechnology research effort.

Public awareness of molecular engineering increased in 1991 as the American Association for the Advancement of Science devoted a special issue of their journal *Science* to the field, with articles on reverse engineering biological systems, molecular self-assembly, atomic and molecular manipulation, and investments in the fledgling technology in the United States and Japan. In the same year, the *New York Times* reported on the second United States conference on molecular nanotechnology. Andrew Pollack, science editor for the *Times*, wrote that, "the ability to manipulate matter by its most basic components—molecule by molecule and even atom by atom—while now very crude, might one day allow people to build almost unimaginably small electric circuits and machines, producing, for example, a supercomputer invisible to the naked eye. Some futurists even imagine building tiny robots that could travel through the body performing surgery on damaged cells. Exploring a range of such possibilities, Michael Flynn published a work of science fiction in 1991, entitled *The Nanotech Chronicles*.

In 1992, Drexler, who is perhaps most responsible for promulgating a vision of molecular robots "performing surgery on damaged cells," moved to substantiate his musings with publication of *Nanosystems: Molecular Machinery, Manufacturing*, and *Computing*. In this technical work, Drexler presents several mechanisms—bearings, gears, cams, clutches as well as computational elements—designed to provide the basic components for nanotechnological assemblers. Though we cannot build these structures today, Drexler argues that assemblers will be feasible within the next few decades. While Drexler's work has received mixed reviews, his effort to describe the details of nanotechnological machinery is currently the most thorough articulation of molecular engineering's potential.

Drexler is convinced that molecular assemblers will be able to build large-scale, molecularly precise structures very rapidly (one-kilogram objects in under an hour) and power them with billions of microscopic computers (each smaller than a blood cell) capable of generating more than 10^{16} (10 quadrillion, or 10 million billion) operations per second. For comparison, the current goal for "high-performance computing" in the United States and elsewhere is the construction of a *teraflop* machine, which would generate 10^{12} (one trillion) operations per second, whereas scientific workstations of the mid 1990s are reaching to perform 10^{9} (one billion) operations per second and most desktop computers perform no more than 10^{7} (10 million).

Whether Drexler's particular vision of the future comes to pass the continuing research efforts of thousands of molecular scientists and technologists—regardless of their chosen title—seem destined to produce an exponentially increasing capacity to build molecularly precise structures, and this capacity will lead to the construction of ever smaller and ever more powerful computational and robotic systems.

In 1992, in order to support the efforts of a variety of molecular researchers, the British journal *Nature* held their first nanotechnology conference in Tokyo. IBM's Don Eigler (the man who spelled out their log with xenon atoms), Richard Smalley (codiscoverer of buckyballs), and others made presentations.

The proceedings of the second conference of artificial life, held in Santa Fe in 1990, also appeared in 1992, as did two introductions to the field of artificial life: M. Mitchell Waldrops's *Complexity: The Emerging Science at the Edge of Order and Chaos* and Steven Levy's *Artificial Life: The Quest for a New Creation.*

In the conference volume, J. Doyne Farmer, a physicist in the Complex Systems Group at the Los Alamos National Laboratory and his wife, Alletta Belin, write that "Within fifty to a hundred years, a new class of organisms is likely to emerge. These organisms will be artificial in the sense that they will originally be designed by humans. However, they will reproduce, and will 'evolve' into something other than their original form; they will be 'alive' under any reasonable definition of the world. These organisms will evolve in a fundamentally different manner than contemporary biological organisms, since their reproduction will be under at least partial conscious control, giving it a Lamarckian component. The pace of evolutionary changes consequently will be extremely rapid. The advent of artificial life will be the most significant historical event since the emergence of human beings. The impact on humanity and the biosphere could be enormous, larger than industrial revolution, nuclear weapons, or environmental pollution. We must take steps now to shape the emergence of artificial organisms; they have potential to be either the ugliest terrestrial disaster, or the most beautiful creation of humanity.

Also, in 1992, the Institute for Scientific Information noted that the prefix "*nano-*" was one of the most popular among new journals, including *Nanobiology* and *Nanostructured Materials*.

In 1993, the National Science Foundation (NSF) committed funding for the formation of a National Nanofabrication Users Network, to include Cornell University, Howard University, Pennsylvania State University, Stanford University, and the University of California at Santa Barbara. The NSF's announcement noted that "Nanofabrication is a critical 'enabling technology for a wide variety of disciplines... The network will help the nation remain at the forefront of many burgeoning research areas, a number of which have commercial applications." Sooner than Feynman imagined—in less than thirty-five years—"the very great future" is at our door.

Research Frontiers

Shifting from the chronicles of a genealogy to the necessarily chaotic impressions of current events, we conclude with some snapshots of recent research.

Atomic and Molecular Sensors

Tools that allow us to observe phenomena allow us to construct tools of manipulation as well as more refined tools of observation. Several sensing and imaging technologies have been developed that provide atomic and molecular resolution. Chief among these are the scanning probe microscopes, including *scanning tunneling microscopes* (STM) and *atomic force microscopes* (AFM). "In the 10 years since it first showed up in the laboratory, the scanning tunneling microscope has distinguished itself as a workhorse in the scientific instrument stable. Priced between $50,000 and $500,000, and quickly becoming a $100 million industry, scanning probe microscopes provide an immediate window on the atomic world.

Picosecond time-resolution is being added to the nanometer spatial-resolution of the STM. Researchers at IBM's Thomas J. Watson Research Center report that the "ability to combine the spatial resolution of tunneling microscopy with the time resolution of ultrafast optics yields a powerful tool for the investigation of dynamic phenomena on the atomic scale. This work extends the imaging capacity of the STM from three dimensions to four, allowing observation of molecular interactions in appropriate length and time scales.

The STM's imaging capacity is also being extended to more accurately resolve molecular structures. Because the STM responds to the electron cloud surrounding an atom, and because atoms bound in molecules share electrons in complex ways, it remains difficult to identify molecular species with scanning probe microscopes. However, recent work at IBM's Almaden Research Center on STM image-processing algorithms has made it possible to image molecular structures with much improved clarity. Their work has "brought researches a step closer to using STM to track the reactants, intermediates, and products of chemical reactions."

Graphically demonstrating Louis de Broglie's 1942 theory that all elementary particles behave as both waves and particles, researchers at IBM's Almaden Research Center used an STM to create and image a circular "quantum corral" of forty-eight iron atoms on the surface of a copper crystal. The corral, fourteen nanometers in diameter, induced the copper electrons to dramatically display their wavelike nature by exciting the otherwise planar array of copper electrons into a standing wave pattern. "When electrons are confined to length-scales approaching the de Broglie wavelength, their behaviour is dominated by quantum mechanical effects. In short, the corralled atoms share their electrons to create a standing wave of quantum-mechanical densities that the STM can perceive. The STM was also used to nudge the iron atoms into place.

Buckytubes and Other Nanotubes

Elongated tubes of carbon, discovered in 1991 as a byproduct of buckyballs, became known, naturally enough, as *buckytubes*. The tubes are roughly the diameter of buckyballs, but they can be several microns in length. Although reliable manufacturing techniques have yet to emerge, the structural properties are known to be unprecedented. Richard Smalley claims that "they would be the strongest fibers we could make from anything. Discovered by Sumio Iijima, a physicist at the NEC Fundamental Research Laboratory in Japan, the carbon tubes have already been recommended for many applications. For example, "two concentric buckytubes may form a wire with an insulating outer shell and a conducting inner shell. This is possible because a buckytube's electronic properties vary with its diameter and other dimensions. The difference is not chemical—as in copper insulated with PVC—but entirely geometrical.

While Iijima has used buckytubes to cast metal wires thinner than DNA, others in Japan and the United States have discovered how to build nanotubes out of various *self-assembling* molecules. For example, Akira Harada, a chemist at Osaka University, has assembled tubules 1.5 nm in diameter from cyclodextrin, a glucose derivative.

Reza Ghadiri, a chemist at the Scripps Research Institute in La Jolla, California, has developed a class of nanotubes built from groups of amino acids, or peptides. "For the first time," claims Ghadiri, "tubular structures on the molecular scale can be used in biological setting ... These nanotubes may be able to form molecular channels, self-assembling inside cell membranes and acting like junctions for transferring molecules into and out of, or between, cells... They're like little test tubes in which we can perform reactions or confine the growth of materials placed inside.

Lipids, which consist of long-chain *aliphatic* hydrocarbons (i.e., the carbon atoms form chains rather than rings), and their derivatives have also been used to construct nanotubes. Because of their aliphatic chains, lipids are insoluble in water but soluble in fat solvents, such as ether, chloroform, and benzene. In an article on lipid tubules. Joel Schnur of the Center for Bio/Molecular Science and Engineering at the Naval Research Laboratory in Washington, DC, argues that self-assembling tubules in an area "ripe for technological breakthroughs."

If we can coordinate the expertise of a number of different disciplines—biology, biochemistry, organic chemistry, inorganic chemistry, physics, and materials engineering—Schnur believes that we should soon be able to "design and engineer... molecules to form self-assembling structures optimized for specific applications."

Reverse Engineering Biological Motors

Biochemists and molecular biologists have discovered and described thousands of molecular mechanical structures used by living organisms—the most famous being DNA, which holds the information used to construct protein molecules. One family of protein machines under intense scrutiny is the "motor molecules" that move tiny components about within the cell. The interior of all living cells, from fungi to humans, is laced with dense cobwebs of microtubules that provide a network of 25-nanometer-wide train tracks for hauling molecular equipment from place to place. These molecular cargo trains "play a role in many of the cells' most fundamental activities. They may help orchestrate the dance of the chromosomes as they separate into the two daughter cells during cell division... They guide the migration of the small membrane-bound vesicles that carry the enzymes that synthesize neurotransmitters to the nerve terminals where the transmitters are made and released. And they may shuttle into place the protein filaments needed for the assembly of large internal cellular structures such as the endoplasmic reticulum, which is where many proteins are assembled."

Using "*optical tweezers*"—a device that pins molecular structures in place with beams of light—Steve Block a biophysicist at the Rowland Institute for Science in Cambridge, Massachusetts, has discovered how kinesin molecules move along microtubule fibers (Fig. 1.11). Although it remains unclear exactly how kinesin consumes ATP to turn over its engine, Block's research group has "shown that a single kinesin molecule takes an 8-nanometer step... result dovetails nicely with recent research from other labs showing that kinesin molecules naturally sit on microtubules at 8-nanometer intervals, a gap corresponding to the spacing of some of the smaller proteins of which microtubules are formed."

Biomolecules for Computation

Another *reverse-engineering* effort that is close to producing functional applications is the exploitation of biomolecules as logical units and memory elements in computational devices. Robert R. Birge, Director of the Center for Molecular Electronics at Syracuse University

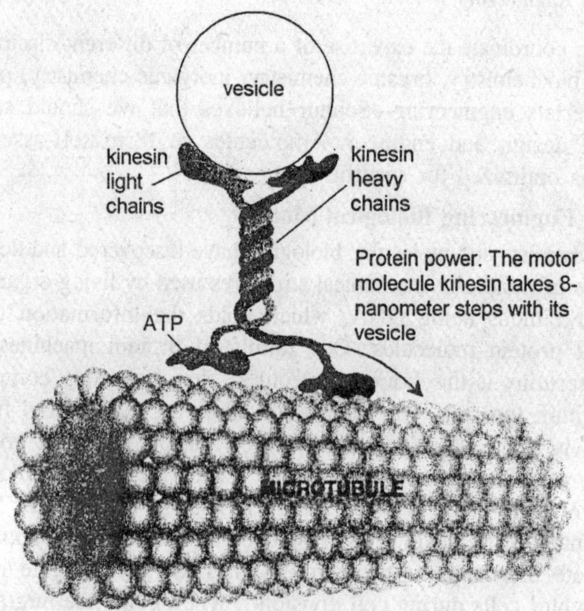

Fig. 7.11. Kinesin, a 60-nm molecule, is a biological engine for intracellular transport.

and an editor for the journal *Nanotechnology*, has developed a technique that harnesses the reaction of the light-harvesting molecule *bacteriorhodopsin*—whose femtosecond response to photons is similar to that of rhodopsin—to store and manipulate digital information. Using two lasers that operate at distinct wavelengths, Birge is able to "read" and "write" data by, respectively, interrogating and flipping the configuration of bacteriorhodopsin molecules. Birge has successfully built three-dimensional, light-addressed memory modules that store 18 gigabytes of data in a block measuring 1.6 cm × 1.5 cm × 2 cm. Birge states that "our current storage capacity is well below the maximum theoretical limit of 512 gigabytes for the same 5-cm^3 volume.

As Birge focuses on bacteriorhodopsin, Jonathan Lindsey, a chemist at Carnegie-Mellon University, is investigating another light-harvesting molecule: *porphyrin*, Porphyrin is a precursor of both the heme molecule (which binds oxygen in animal blood) and chlorophyll. In plants, chlorophyll arrays are extremely complex. Nobody has been able to figure out their structure or precisely how their configuration relates to their light-capturing function. In an effort to study a simplified version of the chlorophyll mechanisms, "Lindsey and his team built porphyrin pentamers, each made of one central porphyrin molecule

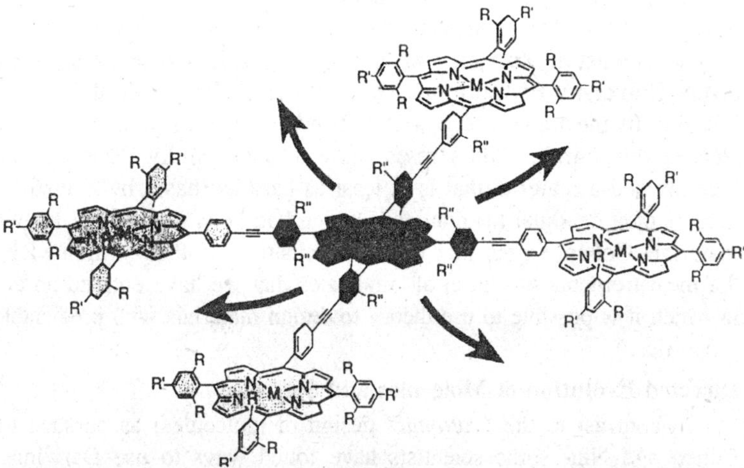

Fig. 7.12. A porphyrin pentamer gathers bluish wavelengths of light at its periphery and emits redish light at its core.

flanked on each of its four sides by another porphyrin molecule. The outer quartet of porphyrins, they found, absorbs light of specific wavelengths and rapidly transfers the light energy to the core porphyrin, which vents the energy input by fluorescing." Although this research is still exploratory, Lindsey anticipates using porphyrin structures to build "molecular information-processing devices" based on the molecular absorption and emission of photons.

Other molecules are also being pressed into service as computational elements. At Queen's University, Belfast, in Northern Ireland, a team led by Prasanna de Silva reports "the fabrication of a single molecule that behaves as an 'and' gate in a logic circuit. The molecule, an anthracene derivative called benzo-15-crown-ether-aldehyde, fluoresces, or emits light of one wavelength, when exposed to light of another. The molecule can function as an 'and' gate because it reacts differently to two inputs: hydrogen ions and sodium ions. The intensity of this molecule's fluorescence varies, depending on whether a signal comes from the hydrogen channel, the sodium channel, or both. When both channels provide input, the molecule radiates at a stronger intensity, clearly signaling that channel 1 and channel 2 are both on. This technology offers the promise that single molecules could replace whole electronic components, such as transistors... In the theory, a cluster of molecules could replace an entire computer chip."

In any attempt to engineer—or reverse engineer—useful molecular structures, a primary hurdle is the ability to predict, form first principles, the atomic machinations and the macroscopic properties of designed, or modified, molecules. Considering the work of Chunming Niu, a chemist at Harvard University, Marvin L. Cohen, a physicist at the University of California, Berkeley, and the Material Sciences Division of the Lawrence Berkeley Laboratory, affirms that we have cleared this hurdle. Niu's research team successfully designed and then created a material that is at least as hard as (has a bulk modulus greater than or equal to) diamond, which had been the hardest known material. Cohen writes that "the confirmation of theory implied by the measurements of Niu et al., indicates that we have entered an era in which it is possible to use theory to design materials with predictable properties."

Directed Evolution of Molecules and Software

In contrast to the "*rational*" design of molecules, as pursued by Cohen and Niu, some scientists have found ways to use Darwinian selection to evolve molecular structures. "Researchers begin with a single molecule, selected for its potential to do some useful chemical task. They make millions or billions of copies of the molecule, each with a slightly varied structure. Then they launch a talent search by making members of this 'population' compete at a task, such as binding to another molecule. They discard those that fail and replicate those that perform well. By repeating this process again and again, each time selecting the best in that generation, scientists... evolve a molecule exquisitely adapted to do exactly what they want." For example, Gerard F. Joyce, a chemist at Scripps Research Laboratory in L Jolla, California, and others used directed evolution to produce an RNA enzyme, or *ribozyme*, that would cut strands of DNA, something it does not naturally do. Using a technique developed in the mid-1980s, *polymerase chain reaction* (PCR), Joyce's team generated copies of the original strand of RNA while introducing mutations into each generation. They selected the best performers from each generation and, after twenty-seven generations, their hothouse ribozyme was 100,000 times more effective than the original. Because PCR is able to copy only strands of DNA and RNA nucleotides, this method does not provide a general purpose technique for evolving molecules, but it clearly demonstrates the power of directed evolution.

Darwinian mechanisms are also at the forefront of software design. Stephanie Forrest, a computer scientist at the University of New Mexico, Albuquerque, writes that "genetic algorithms are a search method that can be used for both solving problems and modeling evolutionary systems. With various mapping techniques and an appropriate measure of fitness, a genetic algorithm can be tailored to evolve a solution for many types of problems." *Genetic algorithms* operate by creating an initial population of binary strings in a computer memory; then the strings are tested for their capacity to execute a given function. Depending on the ability of an individual to perform, it is allowed to "*procreate*" with other successful individuals and produce offspring that must face similar evaluation. Darwinian functions such as variation, selection, and inheritance have been shown to be useful techniques for evolving binary-string representations of functional programs. Indeed, "biological mechanisms of all kinds are being incorporated into computational systems, including viruses, parasites, and immune systems." John Holland, professor of psychology and computer science and engineering at the University of Michigan, and Maxwell Professor at the Santa Fe Institute, points out that genetic algorithms are particularly useful for modeling many "systems of crucial interest to humankind that have so far defied accurate simulation by computer economies, ecologies, immune systems, developing embryos, and the brain."

Central to evolution-based simulations as well as directed molecular evolution is the problem of replication and, in particular, self-directed replication. Given the complexity of known biological replicators, researchers have imagined that self-replication was inescapably complex. As a result, early theoretical work in this area did not strive for models that might lead to physical realizations. Von Neumann's initial proposal for a self-replicating machine, for example, embedded a general-purpose computer into a two-dimensional *Cellular automata* that required 29-state cells. Recently, however, a team lead by James Reggia, a computer scientist at the University of Maryland, has shown that "self-replication is not an inherently complex phenomenon but rather an emergent property arising from local interactions in system that can be much simpler than in generally believed." Extending the work of Chris Langton, who has shown how to build 86-cell, self-replicating, "Q-shaped sheathed loop" cellular automata, Reggia's team has demonstrated a much smaller, but still Q-shaped, two-dimensional structure that requires only five cells. Reggia notes that, "the existence of these systems raises the question of whether contemporary techniques being developed by organic chemists studying autocatalytic systems [in which structural molecules function as templates for their own replication]... could be used to realize self-replicating molecular structures patterned after the information processing occurring in unsheathed loops.

Molecular Computation

Another challenge facing molecular engineers is the design of new computational architectures. For example, one of the most difficult problems in contemporary high performance computing is heat-extraction. The computational density implicit in molecular machinery will require novel architectures if the machines are to operate at acceptable temperatures.

One proposal is to make the operations in a computer reversible. It would then be possible to design systems that could "harvest" information from the tidal flow of instructions through computational ecologies. This approach would avoid, at least in theory, the thermally costly move of intentionally destroying stored information. Over twenty years ago, in 1973, a computer scientist at IBM, Charles Bennett, presented a description of a reversible computer: "In the first stage of its computation the logically reversible automation parallels the corresponding irreversible automation, except it saves all intermediate results, thereby avoiding the irreversible operation of erasure. The second stage consists of printing out the desired output. The third stage then reversibly disposes of all the undesired intermediate results by retracing the steps of the first stage in backward order (a process which is only possible because the first stage has been carried out reversibly), thereby restoring the machine (except for the now-written output tape) to its original conditions". Reversible computation has not been pursued with microtechnological systems, primarily due to the memory requirements for storing intermediate data, but for molecular computing it may well be essential. Indeed, Ralph Merkle, head of the Computatonal Nanotechnology Project at Xerox PARC, claims that "reversible logic will dominate in the twenty-first century."

Exploring another novel approach to computation, John Ross, a chemist at Stanford University, and his team have "shown that a network of chemical-filled beakers can perform one of computing's hardest tasks—recognizing a pattern." That is, groups of interconnected beakers, organized into a *neural net*, can perform logical operations by exploiting various chemical equilibria established among the containers.

In related work, Jean-Pierre Banatre and Daniel Le Metayer, computer scientists at the National Institute for Applied Sciences in Rennes, France, have developed a programming technique based on unordered collections of data objects, such as nested containers of solution-based molecules. This technique is based on the logical notion of a *multiset*, which is the same as a set except that multisets can occurrences of the same element. Banatre and Le Metayer suggest that "an intuitive way of describing the meaning of a GAMMA [General Abstract Model for Multiset Manipulation] program is a metaphor of the chemical reactions: the set can be seen as a chemical solution, function *R* (called the reaction condition) is a property to be satisfied by reacting elements, and *A* (the action) describes the product of the reaction. The computation terminates when a stable state is reached, that is to say, when no elements of the set satisfy the reaction condition." While demonstrating that machinations of multiset transformation provide a powerful technique for describing the operations of parallel-processing systems, Banatre and Le Metayer emphasize the effectiveness and completeness of their model, rather than any practical realization of it. Indeed, they claim they are "not concerned with implementation issue." In contrast, Jack LeTourneau, Chief Scientist at Prime Arithmetics, Inc., has developed an efficient implementation strategy for this kind of "logical parallelism" based on an arithmetic interpretation of finite multisets.

Artificial Cellularization

The notion of hierarchically nested containers is the conceptual foundation of both object-oriented programming and cellular life. And while multisets provide a *logical* tool for describing cellular interactions, recent advances in polymer chemistry are beginning to provide the *physical* tools for encapsulating microscopic molecular systems. Common polymers, such as nylon and polyethylene as well as DNA, consist of linear chains of simple molecules, or *monomers*, which have at least two reactive sites available for bonding. Samuel Stupp, chemist at university of Illinois, has now created *two-dimensional* polymers. In a 21-step process, rodlike molecules with two reactive sites, one at the end one in the middle, self-assemble into 10-nm thick films. "It is perhaps easiest to understand how these precursors are assembled if one imagines that they are sharpened pencils. The eraser corresponds to the reactive end, and the brand name stamped on the pencil represents the central reactive site. The 'brand name' encourages the pencils to align side by side in the same direction. The pencils therefore form a layer with the erasers on one side and the points on the other.

Such two-dimensional polymer films, which easily extend over several square microns, should be able to provide semipermeable

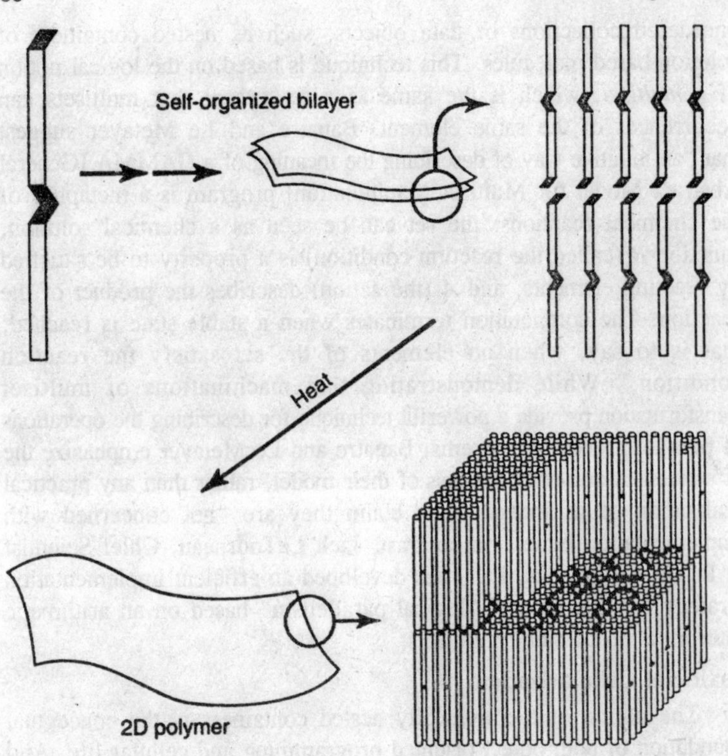

Fig. 7.13. Two-dimensional polymer sheets made from rodlike precursors with two reactive sites.

membranes for encapsulated and connected populations of evolving molecular components. Here we might listen to Harold Morowitz, professor of biology at George Mason University, who has argued that "it is the closure of an amphiphilic bilayer membrane into a vesicle that represents a discrete transition from nonlife to life... It seems likely that … the emergence of deoxyribonucleic acids, transcription, and other elaborations came later... The first radiation leads from the earliest vesicles to the universal ancestors. At this stage biogenesis is over, and the origin of species begins. The rest is history.

8

MACHINE-PHASE NANOTECHNOLOGY

In 1959 physicist Richard Feynman gave exploring statement about the limits of miniaturization. He set out from known technology (at a time when an adding machine could barely fit in your pocket), surveyed the limits set by physical law and ended by arguing the possibility—even inevitability—of "*atom by atom*" construction.

What at the time seemed absurdly ambitious, even bizarre, has recently become a widely shared goal. Decades of technological progress have shrunk microelectronics to the threshold of the molecular scale, while scientific progress at the molecular level—especially on the molecular machinery of living systems—has now made clear to many what was envisioned by a sole genius so long ago.

Inspired by molecular biology, studies of advanced nanotechnologies have focused on bottom-up construction, in which molecular machines assemble molecular building blocks to form products, including new molecular machines. Biology shows us that molecular machine systems and their products can be made cheaply and in vast quantities.

Stepping beyond the biological analogy, it would be a natural goal to be able to put every atom in a selected place (where it would serve as part of some active or structural component) with no extra molecules on the loose to jam the works. Such a system would not be a liquid or gas, as no molecules would move randomly, nor would it be a solid, in which molecules are fixed in place. Instead this new machine-phase matter would exhibit the molecular movement seen today only in liquids and gases as well as the mechanical strength typically associated with solids. Its volume would be filled with active machinery.

The ability to construct objects with molecular precision will revolutionize manufacturing, permitting materials proper ties and device performance to be greatly improved. In addition, when a production process maintains control of each *atom*, there is no reason to dump toxic left overs into the air or water. Improved manufacturing would also drive down the cost of *solar cells* and *energy storage systems*, cutting demand for coal and petroleum, further reducing pollution. Such advances raise hope that those in the developing world will be able to reach First World living standards without causing environmental disaster.

Low-cost, lightweight, extremely strong materials would make transportation far more energy efficient and—finally—make space transportation economical. The old dreams of expanding the biosphere beyond our one vulnerable planet suddenly look feasible once more.

Perhaps the most exciting goal is the molecular repair of the human body. Medical nanorobots are envisioned that could destroy viruses and cancer cells, repair damaged structures, remove accumulated wastes from the brain and bring the body back to a state of youthful health.

Another surprising medical application would be the eventual ability to repair and revive those few pioneers now in suspended animation (currently regarded as legally deceased), even those who have been preserved using the crude cryogenic storage technology available since the 1960s. Today's vitrification techniques—which prevent the formation of damaging ice crystal—should make repair easier, but even the original process appears to preserve brain structure well enough to enable restoration.

Those researchers most familiar with the field of molecular nanotechnology see the technology base underpinning such capabilities as perhaps one to three decades off, At the moment, work focuses on the earliest stages: finding out how to build larger structures with atomic precision, learning to design molecular machines and identifying intermediate goals with high payoff.

To understand the potential of molecular manufacturing technology, it helps to look at the macroscale machine systems used now in industry. Picture a *robotic arm* that reaches over to a *conveyor belt*, picks up a loaded tool, applies the tool to a workpiece under construction, replaces the empty tool on the belt, picks up the next loaded tool, and so on—as in today's automated factories.

Now mentally shrink this entire mechanism, including the conveyor belt, to the molecular level to form an image of a nanoscale construction system. Given a sufficient variety of tools, this system would be a general-purpose building device, nicknamed an assembler. In principle, it could build almost anything, including copies of itself.

Molecular nanotechnology as a field does not depend on the feasibility of this particular proposal—a collection of less general building devices could carry out the functions mentioned above. But because the assembler concept is still controversial, it's worth mentioning the objections being raised.

One prominent chemist speaking asked how one could power and direct an assembler and whether it could really break and re-form strong molecular bonds, These are reason able questions that can be answered only by describing designs and calculations too bulky to fit in this essay Fortunately technical literature providing seemingly adequate answers has been available since at least 1992. Another well-known chemist objects that an assembler would need 10 robotic "*fingers*" to carry out its operations and that there isn't room for them all. The need for such a large number of manipulators, however, has never been established or even seriously argued. In contrast, the designs that have received (and survived) the most peer review use one tool at a time and grip their tools without using any fingers at all.

These examples point to the difficulty of finding appropriate critiques of nanotechnology designs. Many researchers whose work seems relevant are actually the wrong experts—they are excellent in their discipline but have little expertise in systems engineering. The shortage of molecular systems engineers will probably be a limiting factor in the speed with which nano technology can be developed.

It is important that critiques of nanotechnology are well executed, because vital societal decisions depend on them. If *molecular nanotechnology* as described here is correct, policy issues can look quite different from what is generally expected. Today most people believe that global warming will be hard to correct—with nanotechnology, excess greenhouse gases could be inexpensively removed from the atmosphere. Current Social Security projections assume increasing numbers of aged citizens in poor health. With advanced medical nanotechnology tomorrow's seniors could be more active and healthy than they are now, bringing new meaning to the "*golden years.*"

Likewise, focusing on avoiding accidents and preventing abuse of this powerful technology Solid work has been done on the problem of heading off major nano technology accidents.

But the challenge of preventing abuse—the exploitation of this technology by aggressive governments, terrorist groups or even individuals for their own purposes—still looms large. The closest analogy to this problem these days is the difficulty of controlling the proliferation of chemical and biological weapons. The advance toward molecular nanotechnology high lights the urgency in finding effective ways to manage emerging technologies that are powerful, valuable and open to misuse.

9

Designing Molecular Components

A *proto assembler* could be described as the "stupidest thing you could imagine, but able to make something better than itself." It will use a considerable amount of direction from the "*outside*." We apparently have two paths leading toward a proto assembler, but each path has a major problem associated with it.

The first path is the *biological path* which is very promising because it is possible to do a tremendous number of things with biological systems right now. In fact, it is present possible to make practically any desired biological molecule. Unfortunately, it is not possible to design a biological molecule for a novel purpose. Generally, you cannot write down the specifications for a reaction to be catalyzed by a novel enzyme, and then derive the structure of the enzyme. We do not know how it designs a biological molecule to perform a specific function.

The second path is the *mechanical path* that Eric Drexler designed the machines. Progress along either of these two paths would be very welcome. One opportunity for progress along the biological path is through the promising idea of *artificial evolution*. It is now possible to evolve an enzyme that will perform a predetermined function. No one knows why the molecules produced through artificial evolution work, because the molecules were produced not by design, but instead by imposing conditions for Darwinian evolution in the test tube.

For progress along the mechanical path, having molecular building blocks that could be assembled with an *atomic force microscope* (AFM) would be very useful. In this way, structures that had already been designed could be assembled in fewer, and more standardized, steps. In contrast, adding individual atoms or small groups of atoms would present unique problems with each step of the construction.

Artificial Evolution

It will be describing the work of Dr. Gerald Joyce of The Scripps Research Institute in La Jolla, CA. The work was published in *Science* on July 31, 1992. Fig. 9.1 shows the *Tetrahymena ribozyme*. This is a piece of RNA that cuts itself out of a longer piece of RNA, so it is what is called a *self-splicing intron*. The intron grabs both ends where it is attached to the longer piece of RNA, folds the RNA strand into a specific shape, and then cuts itself out of the longer RNA strand. The existence of such ribozymes is part of the evidence that leads us to believe that there was once an "*RNA world*." All biological molecules were made of RNA—both the information carrying molecules and the molecules that did things (the *enzymes*). We are still discovering remnants of the role of RNA as an enzyme, even though most enzymatic

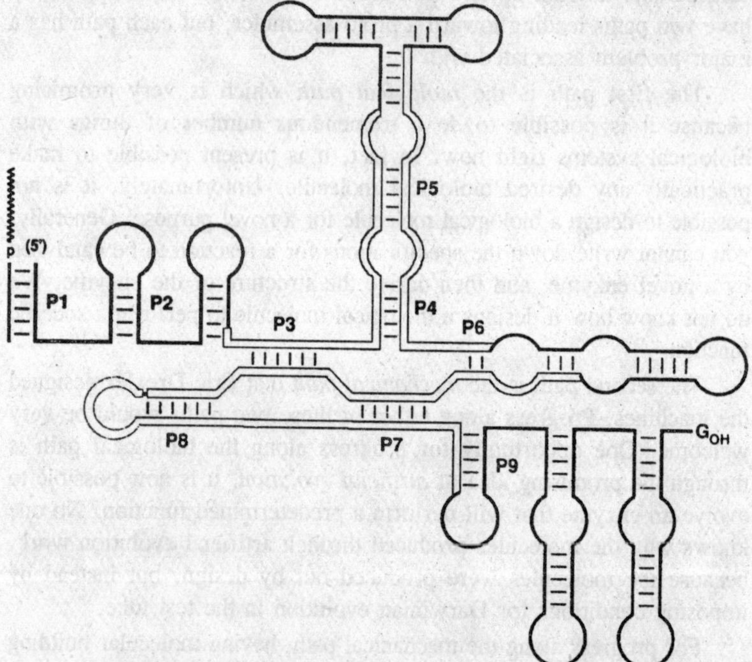

Fig. 9.1. Secondary structure of the Tetrahymena ribozyme (L-21 form) showing those regions that were randomly mutagenized (boxed segments).

function has been taken over by proteins, which do a better job as enzymes.

The *Tetrahymena ribozyme* is able to cut and splice RNA. The challenge is to make an RNA enzyme that can cut and splice DNA. We are not aware of any such ribozymes existing in nature, although some ribozyme mutants can cleave DNA very inefficiently. To make such a ribozyme, Beaudry and Joyce developed a process that, in a single test tube and at a constant temperature, is able to reproduce RNA. However, to amplify the amount of RNA that is produced, they also use the PCR (*polymerase chain reaction*) technique to make many copies of the intermediate c DNA molecule that is formed by their process, and that is then transcribed again to make more RNA molecules.

The technique thus developed can make many copies from a very small amount of RNA. Theoretically, you could start with a single molecule of RNA in a test tube, add the proper components, come back later, and have a million copies of that RNA molecule.

An important component of this process is the deliberate choice of conditions to make the copies contain errors. After the million copies have been made (with modified PCR), many variations of the original sequence will have been introduced into the copies. After a number of cycles of amplification, each one introducing additional variations in the sequence, an effective DNA-cleaving enzyme was obtained.

There is no natural RNA enzyme with this activity. This new ribozyme might have some practical uses. Perhaps it could be used to inoculate individual cells against certain viruses. If the enzyme is in the cell, and the virus enters the cell, the enzyme recognizes the viral DNA and chops it up.

The enzyme that was produced is what Richard Dawkins calls *designoid*. A designoid object is something that looks like it was designed because it seems to have purposeful parts, but was instead produced by evolution. Many of the things in this room, including the people, are designoid in that no designer actually designed them, but they nevertheless came out rather well. Other things here are designed by conscious minds in the usual manner.

The actual process that was used in this artificial evolution is shown in Fig. 9.2. The RNA enzyme combines with a small piece of DNA to form a complex. The RNA then does its magic and cuts the DNA, leaving only a shorter piece of DNA complexed to the RNA.

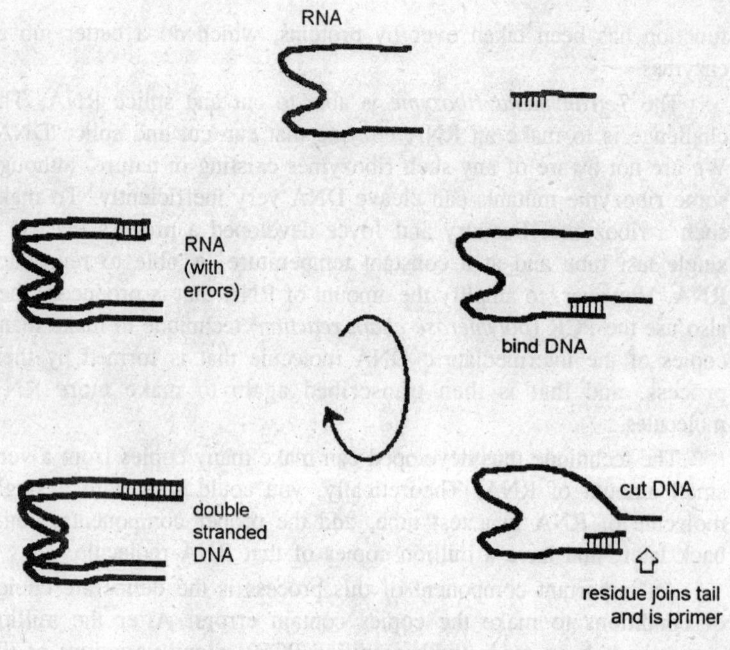

Fig. 9.2. Process of artificial evolution of RNA enzyme in the test tube.

The ribozyme splices the piece of DNA (that has been cut off) onto its own tail. The little piece that has been spliced onto the tail of the RNA is now the priming sequence for copying the RNA. That small piece of DNA spliced onto the tail of the RNA molecule is essential for that specific RNA molecule to be copied. Therefore, only those RNA molecules that are able to cut the DNA, and join the end of the DNA to the RNA, will be able to start the copying process. During the copying process this RNA-DNA complex is converted into cDNA (that is, a DNA copy of thc original RNA), then into double-strand DNA, and finally into RNA, returning to the beginning of the cycle.

Steps in Artificial Evolution

The *key step* in this process was to hook the goal property (the ability to cut DNA) into the cycle for replicating the RNA molecule.

Thus, the RNA could not be replicated unless it performed the desired function. Beaudry and Joyce had to use some tricks to get this cycle to work, because the original RNA molecule was not capable of cutting DNA. They began their experiment not with the original ribozyme, but with billions of variants of the molecule made artificially. They chemically synthesized pieces of DNA corresponding to certain portions of the molecule. With a small percentage of random errors, they incorporated these pieces into a gene for the ribozyme, and then transcribed that gene to make mutant RNAs.

Fig 9.3 shows another schematic representation of the ribozyme, laid out in a horizontal plane. The vertical bars represent the number of mutations that occurred at each of the 413 positions in the RNA molecule. To begin the evolutionary process, they made equal numbers of mutations in many parts of the molecule, as shown in Panel A (certain regions could not be mutated without preventing replication, such as the sites for binding the DNA primer). After three cycles of selection and amplification, it is apparent that mutations in certain sites are much more heavily represented because they are the mutations that lead to better function—that is, better ability to cut DNA. After six cycles and after nine cycles, the mutations that serve to turn this ribozyme into one that will cut DNA, are quite apparent. This graph also makes clear what portions of the RNA molecule are involved in the actual cutting function. This process also provided new scientific information about how this ribozyme works. I would also like to point out that the three-dimensional structure of this RNA enzyme is not yet known. Therefore, new capabilities were created without knowing the structure of the molecule.

Consider the following important features of this artificial evolution experiment:

1. A test tube amplification system amplifies the input of RNA molecule a million-fold, but only for those molecules that perform the desired function.
2. The function that is desired must be made essential for the replication of molecule. Joyce and his colleagues have also been able to use artificial evolution to change the ribozyme's requirement for magnesium in solution to a requirement for calcium. However, he was not able to evolve the ribozyme to dephosphorylate glucose. This was too large a step to take. Perhaps if you approach this goal gradually, you could get the molecule to adapt to perform that function as well.

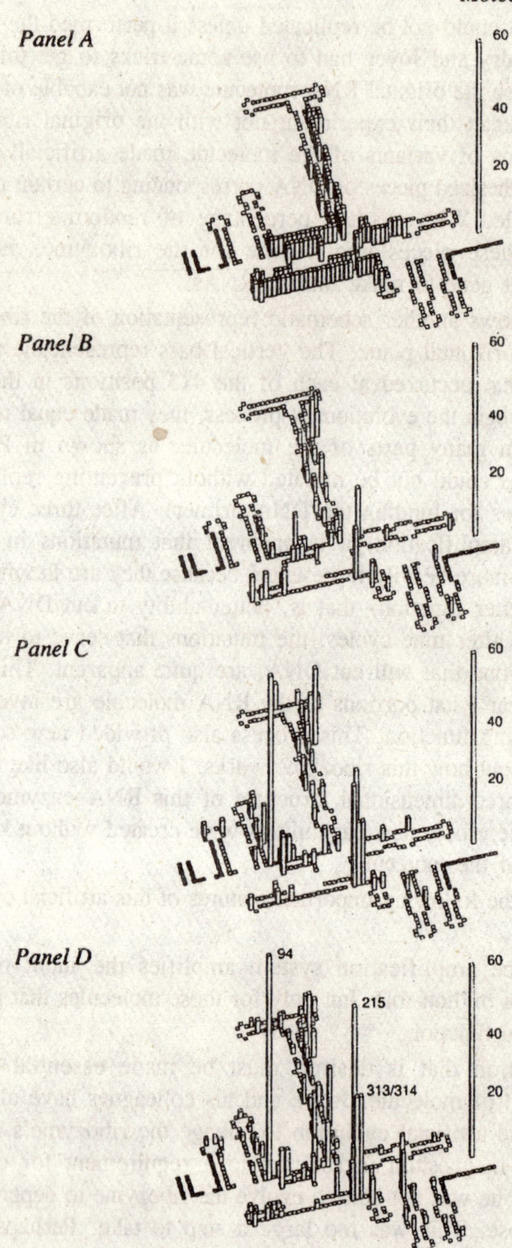

Fig. 9.3. Sites at which mutations occurred over the course of the evolution of the ribozyme.

3. The introduction of errors, which are a very important part of this process, was done in two ways, both outside of the steps done in a single test tube. First, the initial population of RNA molecules had been altered by transcribing them from mutated genes. The mutated genes had been constructed by using synthetic DNA that had been randomly mutagenized. Second, the PCR reaction used as part of the amplification step for each generation introduced additional errors in each generation.

Applications of Artificial Evolution

Several companies are exploiting this artificial evolution technology, mostly to make drug molecules. Many of these companies are synthesizing small chains of peptides or nucleotides to block some biological process, or bind to some biological receptor. Several academic groups are also working with this technology.

What are the long-term goals of this sort of research? The "*Holy Grail*" here is to be able to evolve proteins, because proteins are very good enzymes, and we currently do not know how to design them effectively. If we could apply the artificial evolution process to proteins, we could make really effective new catalysts. The problem is that once you perform the selection step and find something that works, how do you make more of it? No process is known to go backward from a protein to DNA.

One trick to accomplish this step might be to attach the RNA that encodes a protein to the end of that protein molecule. Or, perhaps it would be possible to "*freeze*" or fixate the ribosome as it is translating the last portion of the protein, and to test the entire complex of protein, ribosome, and messenger RNA for activity, and then isolate the complexes that work. Then it should be possible to isolate the messenger RNA from the complex, copy it into DNA, and amplify it while introducing additional variations.

What does all of this have to do with nanotechnology? We must make parts in order to make molecular machines. We currently can make a biological molecule, protein, RNA, or DNA, of practically any sequence that we wish. It is even possible to order small DNA molecules commercially with 48-hour turnaround times.

The trouble is that we do not know what sequence to write down that would make a good part for a nanomachine. Suppose that we could somehow characterize what that part was supposed to do in a way that made those characteristics a requirement for reproducing that molecule. Then we could evolve a molecule that could do what we want, without even knowing in advance the shape of the molecule. The tricks for hooking the desired characteristics into the *molecular reproduction* process are going to become more sophisticated because much effort is being invested in this work.

There are reasons to be optimistic about the contribution of artificial evolution to the development of molecular nanotechnology. First, there has been extremely rapid progress along this path. Second, companies have already been formed to use this technique for drug design. Third, as soon as these companies become profitable, the snowball will start rolling, making the development of this technology self-financing, like the microelectronics industry has been since 1955.

Molecular Building Blocks

The second opportunity for progress toward a proto-assembler lies along the mechanical path to a proto-assembler. This path is also called the *mechanosynthesis path*. Markus Krummenacker of the Institute for Molecular Manufacturing has been doing some very interesting work designing molecular building blocks.

Suppose that we had a modified AFM tip that could be precisely positioned over something that we are trying to build. Such a system could be used to build "*by hand*" a more advanced proto-assembler. Given such a system, what bound ligand would we like to deposit on the product structure in order to build a useful device? Construction of a proto-assembler might be possible soon if we could assemble it from well-defined building blocks placed in a three-dimensional lattice.

Krummenacker is looking for molecular building blocks that could be used to build more complex structures. These building blocks would be designed molecules, synthesized by using conventional chemical techniques, that could be placed on a work piece, and made to covalently bind to that work piece structure.

Requirements for Molecular Building Blocks

The requirements for such molecular building blocks are as follows:

1. It must be possible to make the building blocks by conventional chemistry, because they are needed before molecular manufacturing is available.
2. It must be possible to physically place the building blocks with an AFM.
3. Once the building blocks are placed on the product structure, they must bond covalently to that structure to irreversibly form a strong structure.

4. The building blocks must be sufficiently general (a cube, for example) so that they can be assembled in a variety shapes. They have to fulfill certain geometric lattice construction constraints.
5. The building blocks must have places to attach chemically active functional groups. Once you have finished building a structure, you want to have not just a chemically inert structure, but one that does something.

Problems in Finding Molecular Building Blocks

Two basic problems must be faced in searching for such molecular building blocks. What should the basic structural skeleton be? And what chemical reaction should be used to covalently attach the building block to the work piece under construction?

Several specific problems are encountered when considering what should be used as the structural skeleton. Three-dimensional organic molecules are surprisingly difficult to find. Most organic molecules are *flat* (like benzene), *floppy* (like the molecules in liquid crystal displays), or *long polymers*. None of these molecules have a stiff three-dimensional structure.

A suitable structural skeleton must also have at least six places to attach (at the top, bottom, left, right, front, and back), so that the pieces can fit together into a lattice. Additional bonding sites also must be present so that functional groups can be attached. Furthermore, the structure must be reasonably stiff. The molecule must be compact and not too leggy, floppy, or rickety, so that the formal structure will be compact. It would also be nice if the molecule were not completely symmetrical, so that all of the corners did not look the same chemically. Such differences would allow conventional chemical reactions to attach specific connectors or functional groups to specific sites on the building block. If all the corners and all the edge looked exactly the same (like a buckyball), then reactions could not be target to specific positions on the building block.

Krummenacker has not yet been able to find a skeletal molecule that satisfies all of these constraints, although he does have some strong candidates. The following thoughts indicate the direction in which this project is going. Factor to be considered here are the skeleton molecule itself, the bonds that link it wit other skeleton molecules, and the lattice that results from linking these molecules together.

One problem is that the links connecting two building blocks are actually fairly long, leading to a lattice that has many empty spaces

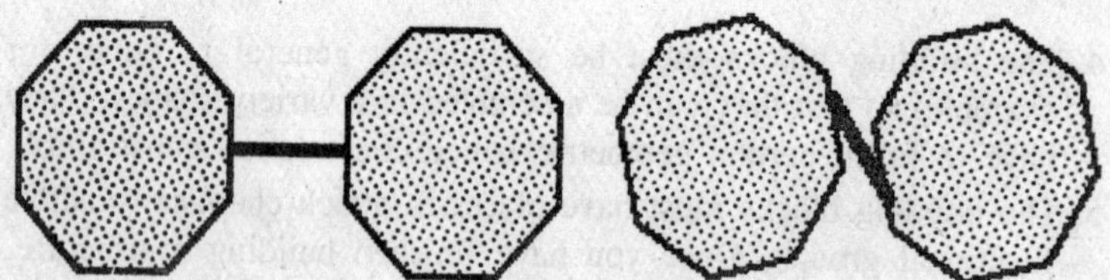

Fig. 9.4. Bond geometry to give a compact lattice assembled from molecular building blocks.

and is not very strong. What is desired instead is that the lattice should be compact and stiff. An apparent solution to this problem is to fold the links, but instead have them come out of the structure at an angle. When blocks are linked together in this fashion, they are then closer to each other than the length of the link between them.

The other half of the problem is finding a chemical reaction that forms the bonds to join the molecular building blocks. This is a potentially very difficult problem. Krummenacker has been considering the Diels-Alder reaction to produce these bonds. The reactants are an open diene connected to some structure, and a dienophile connected to some structure. These reactants have the property that when pressed together, they will rearrange into the configuration. All of the four side groups on the reactants remain present. No atoms have been lost and there are no leaving groups. Furthermore, two bonds are formed between the reactants, making the resulting structure quite stable.

The requirements for a good reaction to bond together molecular building blocks are as follows:

1. The reactants must react within a reasonable time frame when pressed into place.
2. The reactants must not react while they are in storage. It is difficult to satisfy the requirement that the reactants only react when placed on the structure being built. One potential solution to this problem is to use an antibody molecule to grab the reactant molecule from dilute solution. Because of the very high affinity of the antibody for the reactant, the reactant can be stored at very low concentration in the solution and still be bound by the antibody molecule Because the reactant is stored at very low concentration, there is little opportunity for molecules of reactant to come into contact with each other and to react prematurely.
3. The reaction must not generate by-products. Many reactions, for example, produce water as a by-product of bonding together the two reactants. By-products are undesirable because they can get

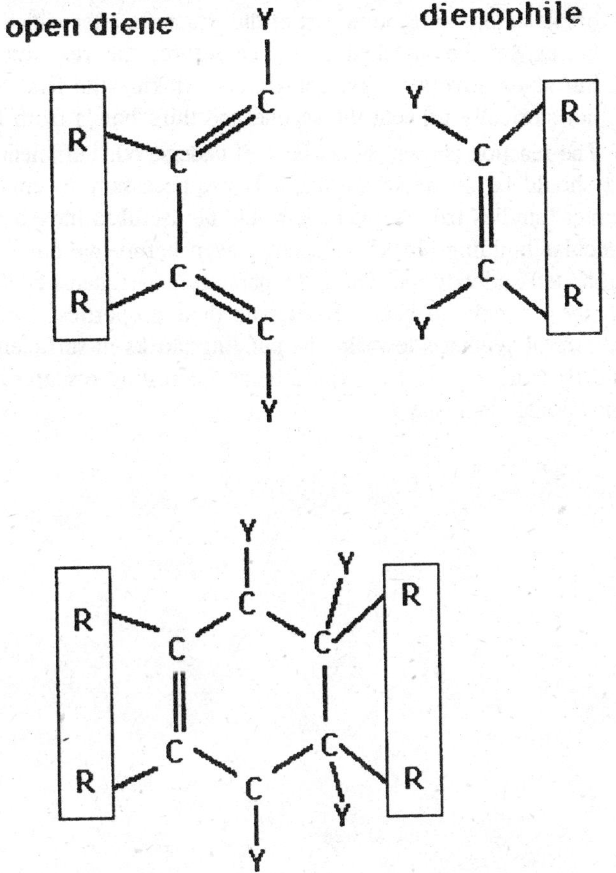

Fig. 9.5. A proposed bond for mechanosynthesis: the Diels-Alder reaction.

in the way of other things. One must plan for an exit pathway for the by-products out of a crowded environment.

4. The reaction must be water-compatible. The proposal is to use an AFM tip with an antibody molecule, and the antibody molecule must be in an aqueous environment to work.
5. The reaction's geometry must be such that a new building block can be bonded in three places, in succession, to build a three-dimensional lattice. For example, consider building a structure from bricks. The brick that is being added attaches to three other

bricks before it becomes part of the structure, after which additional bricks can be attached to it. Therefore, the reaction must not introduce adverse steric constraints. Making the first bond must not sterically prevent the second and third bonds from forming.

The reaction chosen should be well understood. Sufficient empirical data should be available so that it is not necessary to conduct basic chemical studies from scratch. It would be useful to have a design for molecular building blocks right now, even before we have AFM tips with bound antibody molecules. Preparatory experiments could be done with these blocks to better understand their properties. The methods of chemical synthesis to make the building blocks in sufficient quantity could be worked out. This would be an interesting research project in conventional chemistry.

10

Cosmetic Nanosurgery

One of the earliest and most rudimentary applications for nanotechnology may well be in what many might consider a frivolous cause—the alteration and enhancement of human appearance, otherwise known as the *beauty business*. Why? Because the technology can be relatively simple, there is a great demonstrated demand, and vast sums of money can be made. Even with the crude techniques available today, the *cosmetics industry* is thriving and lucrative. Once the means are at hand to actually perform safely, painlessly, and inexpensively the miracles that advertisements now promise, the race will be on.

Market Potential

How big the beauty business? That depends on how define it, but a few representative figures will provide a rough idea. According to various sources, the worldwide gross volume for toiletries in 1990 was in the range $14 to $18 billion. That's just for traditional personal hygine products such as powers, sprays, perfumes, and deodorants.

The appearance enhancement market goes far beyond that. The diet industry is said to gross $33 billion annually. A recent survey by *Glamour* magazine found representative young American women spending from $550 to $400 a year on their looks, including such items as contact lenses, gym sessions, and bicycles for exercise. And, according to a recent *Wall Street Journal* article, the total spent for breast implants, hair transplants, facelifts, tummy tucks, and other surgical repair operations comes to about one billion dollars a year, most of it for women but a growing percentage (about 30 per cent by some estimates) for men.

These expenditures are for goods and services that, by and large, perform far less miraculously than their advertisements would lead one to believe. Indeed, beauty-related goods and services are named in ways that suggest a level of performance unattainable in practice, such as "permanent" waves that last a few months at most.

Far more pervasive, and insidious, are adds that use *gorgeous models* to imply—without really saying—that using the product will make you look beautiful too. Surely no one consciously believes such advertising, but the implication is there. The subconscious naively accepts and reacts.

No one knows what people would pay for products that really did transform one's appearance. It's reasonable to suppose, however, that it would at least equal and perhaps far surpass what they now pay for cover-ups and nostrums.

And this is not the full extent of the market for *cosmetic nanosurgery*. Once this technology begins to mature, it become possible to address beauty needs that previous generations simply bore without hope or recourse. Typical examples might include completely scarless facial surgery, rebuilding poorly set bones to normal contours, and repairing nerve damage that affects facial muscles.

Applications of Cosmetic Nanosurgery

Hair Colour

To get an idea of cosmetic nanosurgery's potential, consider a hair-colour product as a simple example. This might take the form of a rudimentary, *nanomachine*—little more than a carefully designed molecule—that simply circulates in the bloodstream until it finds itself in a hair follicle in the scale. Then it inserts itself into one of the follicle's melanocytes, the cells deep within the hair-root bulb that colour the hair, where it benignly but persistently but regulates the production of delivery of the pigment melanin.

Hair colour, except in the case of *gray hair*, depends less on the amount of melanin than on its distribution. *Red hair* has regularly spaced melanin granules, whereas in brown hair the melanin granules are irregularly spaced. *Black hair* contains large melanin granules, an in *blond hair* they are relatively small.

While the effect of a single melanocyte changing to a different mode of melanin production would be detectable only with a microscope, a small syringe could easily contain more such nanomachines than there are melanocytes on your whole scalp. As a result, a brunette could bleach her hair once and never again have to worry about dark roots growing out. She would, in effect, become a blond from the inside out, requiring only an occasional booster shot to replace any nanomachines that get trapped accidentally in the hair shaft and dragged, out of the follicle.

Further, just as in normally *blond hair*, no two strands of hair would have exactly the same shade. (That's how you can tell a bleach job from the real thing). The colour of each hair shaft would depend on how completely the melanocytes had been converted to the new mode, which would naturally vary somewhat from follicle to follicle. And, since hair waviness is also controlled by what happens in the follicle, it might be subject to similar modification with a suitable nanomachine.

Such a hair-colour nanomachine could be rudimentary. It would require no on-board computer, just the ability to recognize first a scalp follicle and then a melanocyte. It could do both by binding temporarily to an appropriate surface protein on the target cell. Then it would simply bind to some structure within the melanocyte in such a way as to influence melanin formation. There would be no need for the self-replication machinery an assembler would require; in fact, self-replication would be a real disadvantage in this and many other applications.

Some sort of self-replicating machinery would be necessary to produce these *rudimentary nanomachines* in useful numbers, of course. This machinery need not be completely human constructed, however. It should be possible to engineer a number of bacterial strains to produce the various components of these simple nanomachines. These components would then self-assemble when they were mixed together, as virus components do.

A similar nanomachine could be programmed to have the opposite effect—to restart melanin production for those who wish to restore the hair colour of their youth. It would probably need to be more sophisticated than the nanomachine we just described. There are several causes of gray hair, so it would need to be "*smart*" enough to recognize and evaluate symptoms and prescribe a cure. Alternatively, one might design several different varieties of nanomachine, one for each known cause of gray hair, and inject the proper one to treat a particular patient based on the result of testing.

Suppose someone who has taken this treatment changes his or her mind. Suppose gentlemen begin to prefer brunettes, or fashion suddenly decrees that redheads deserve to have more fun. Would the new blonds be stuck with their choice?

Not at all. It should be a simple matter to design a similar hair-follicle-seeking nanomachine to simply turn off the first one. (It would take a different and more sophisticated nanomachine to turn a natural blonde into a brunette.). In fact, both nanomachines could then allow themselves to be incorporated into the growing hair shaft and thus eliminate themselves from the body. This flexibility would also be desirable from a commercial standpoint, giving changing fashions, a chance to provide a continuing market.

It has been suggested that gene therapy, already in clinical trials for certain rare diseases, could probably address thee same needs and be equally reversible. Gene therapy is also likely to be available long before functional nanomachines appear. However, gene therapy will have major political obstacles and formidable buyer resistance to overcome before it becomes accepted for taking care of anything other than life-threatening diseases.

Skin Colour

A similar set of nanomachines, perhaps even simpler because they would have no need to seek but *hair follicles*, would allow people to control their skin colour. These machines could be made to act locally—to deal with *birthmarks*, *liver spots*, and *freckles* for example—or allowed to simply enter whatever melanocyte they encounter anywhere on the body. People could get whatever skin shade—or eye colour—they choose, lighter or darker. The social consequences of such innovations are perhaps best left to the imagination. There can be little doubt, however, as to its market potential.

Baldness

Before leaving the subject of hair follicles, it seems evident that there would be stupendous market for a nanomachine that would restart hair growth in follicles that had turned off. There is probably no branch of the cosmetic industry that has garnered more business with less benefit to the customer than the potions that promise hair restoration. A product that actually worked would be an instant sensation.

Lest anyone argue that hair restoration is an impossible dream, it might be well to mention that not all hair loss is permanent, even now. Chemotherapy and various medical conditions can cause complete hair loss, which reverses itself when the therapy is discontinued or the condition is corrected. And it seems intuitively reasonable that a nanomachine working from inside the hair follicle, right where the action is, should have a better chance of success than any amount of medication trying to fight its way in from outside the scalp.

One problem in developing any such product would be to guarantee that it would function only where one wants it to. No one wants hair sprouting from their forehead or eyelids, for example. (Yes there are hair follicles on your eyelids. They just produce very short and nearly invisible hairs.) Similarly, women with thinning hair (it happens) would not be interested in a hair restorer that also grows beards. The product would have to include strong safeguards against indiscriminate action.

To accomplish this, nanomachines would need to have some way of *"knowing"* where they are in the body. However, this need not be nearly as complicated as might be imagined. One simple way might be with subcutaneous injections of a colourless chemical marker—an invisible tattoo—that would gradually seep into the surrounding tissues. This would set up a localized concentration gradient that the nanomachine could recognize as its start-up signal. Once the nanomachine had become attached to a particular follicle it would stay there, indifferent to the gradual dissipation and eventual disappearance of the chemical marker.

Unwanted Hair

Unwanted hair is an ongoing problem for many men and women: hardly anyone really enjoys shaving. A nanomachine that would enter hair follicle cells and stop unwanted hair production once and for all would be welcomed with *enthusiasm*.

An alternative and possibly simpler way to avoid the need to shave, although it might involve much more mature technology, would be with a one-time external application of a *nanomachine-laden depilatory cream*. In this case, the machines would attach themselves permanently to the stubble ends and busily convert the protein molecules of each hair stub (almost all keratin) into harmless, odorless gases such as methane, nitrogen, water vapour, and carbon dioxide, which would unobtrusively dissipate. It should be possible to get them to digest each hair stub faster than it grows out, eliminating shaving forever.

Sulfur, which constitutes four percent to six percent of hair by weight, is not really a troublemaker, even though all of its volatile compounds are smelly (mercaptans, sulfur dioxide, hydrogen sulfide) and some are skin irritants or toxins. Elemental sulfur, on the other hand, is relatively inert. It should be safe to let it accumulate in little crystals until the next washing removed it. The only potentially troublesome nonvolatile constituent of hair protein appears to be phosphorus, which is present in very minor amounts. *Phosphorus* itself can ignite on contact with air, producing skin-irritating oxides and acids. However, it is unnecessary to carry the breakdown of the hair protein that far. It should be simple to program the nanomachine to leave the phosphorus in a harmless water-soluble organic compound that would wash away in the next rinse.

As a further refinement, these nanomachines could be made to depend on a continuous supply of atmospheric oxygen. For example, the energy to do their work could come from flameless oxidation-reduction reactions the nanomachines would be promoting. This might provide a way to regulate their activity; whenever the supply of oxygen diminished, the nanomachines would slow down and wait for it to pick up again. This would be likely to happen whenever the end of the hair stub sank far below the skin surface and became covered with sebum (the natural hair lubricant secreted into each follicle by its sebaceous gland). In any case, some such regulator might be advisable to assure that no overenthusiastic nanomachines could work their way deep into a follicle and somehow damage it.

For this scheme to work, it would be important to give the nanomachines the ability to distinguish between hair keratin and the skin keratin, and keep them from harming the latter. One way to do this might e to give each nanomachine at least five other arms (or attachment sites) in addition to the one for manipulating keratin structure, and require them to link up in a way that would be easy on hair an hard on skin. For instance, we might make the manipulator arm inoperative unless the other arms were linked to two adjacent *cystine molecules*—the predominant amino acid component of keratin—perhaps by recognizing cystine's characteristic disulfide group, and also to at least three other nanomachines. Hair is almost all keratin, whereas skin keratin is mixed in with other components, making it hard for the nanomachines to fulfill the attachment criteria on skin.

Nanomachines that failed to attach themselves in this way would be neutralized with a follow-up rinse. Those already attached to keratin, either on the skin or on a hair stub, would thereby become immune to the rinse.

Any nanomachines that did attach themselves to skin keratin would find slow going in comparison with the others on the hair stubs. They would keep running into nonkeratin components that they would be unequipped to disassemble. Long before they had made much headway, the skin particle they were on would flake off. Once such a nanomachine ran would in all probability come apart before it could attach to another source of keratin.

At the same time, this multiple-attachment requirement would make it possible for each correctly attached nanomachine to stay stuck to the hair shaft even as it cuts away the ground under its own "feet." Whenever one of its attachment sites to the hair shaft became undermined, it would just grope around with the free arm until it found a fresh handhold and then keep digging on the other side.

One obvious precaution would be to prevent this nanomachine from attacking the keratin in fingernails, which appears to be chemically indistinguishable from that in hair. The easy way to do this would be to wear gloves, but that still leaves open the possibility of damage through carelessness. It would undoubtedly be safer to provide a positive deterrent.

The solution might be to make the nanomachines susceptible to attack by some simple solvent such as acetone, a common nailpolish-remover ingredient and something not normally found on the face. This would also make it possible to readjust the boundary of the area within which the nanomachines can act if one wants longer sideburns or decides to grow a mustache.

Permanent Breath Freshener

A similar topical application would address another personal hygiene concern: bad breath originating in the mouth. As most of us are aware, halitosis frequently results from decaying food particles that perfunctory brushing has failed to dislodge. A useful preparation would bond itself firmly to the tooth enamel and prevent anything else from sticking. It would also actively remove anything that became mechanically wedged.

In this case one would start with a thorough professional cleaning to remove any *accumulated plaque* and *tartar*. The dental technician would then paint on a solution teeming with nanomachines that cling tenaciously to every bit of exposed tooth enamel and link up to form a complete, invisible surface film around each tooth. Each of these nanomachines would then go to work disassembling any foreign matter that came in contact with the tooth into harmless and bacteriostatic gases or liquids, which would quickly be removed by saliva.

The same sort of idea could have many other applications such as always-clean windows, self-cleansing dishes, the permanently spotless and odor-free bathroom, and a true antifouling paint for boats. The latter would enjoy a multimillion-dollar market all by itself. For these applications the nanomachines could have much simpler attachment and attack criteria.

To return to the *fresh-breath nanomachine*, such a system would obviously need to be "*smart*" enough to recognize gum and tongue tissue and leave them alone. It might also need to be periodically reapplied, as normal chewing and brushing would tend to remove it from exposed surfaces. However, the protection would still be there on the hard-to-reach surfaces, and that's where it's most needed in any case. As an added bonus, the nanomachines could be programmed to construct oil of peppermint (or whatever aromatic essence might be desired) out of the organic molecules they have been dismantling.

Although such a rudimentary nanomachine system would certainly improve mouth cleanliness and go a long way to prevent cavities and tartar buildup, it would not by itself be enough to prevent periodontal (gum) disease if it were used as an excuse to stop gum massaging and flossing. That advance will probably have to await the development of fully capable cell-repair nanomachines, which would conquer all diseases including gum disease. Until then we will have to keep supplying manually the exercise our gums need to keep them vigorous and healthy, a discipline that is necessary because our usual bland diets of refined foods fail to provide the stimulation. One hopes that the promoters of a breath-freshener system will market it as a between -brushing safeguard instead of a dental hygiene cure-all.

Wrinkle Repair

As cosmetic nanosurgery technology matures, it should become possible to undertake more ambitious transformations such as skin rejuvenation. They saying, "*Beauty is only skin deep*," simply serves to emphasize how very important the skin is.

The skin is a very complex and little appreciated organ, exposed to a wide range of environmental assaults that would simply destroy any of our other vital organs. It resists extremes of heat and cold, dryness and abrasion, biologic, organic, and inorganic poisons, invasive organisms, the force of gravity, and continual bending and flexing. No wonder it begins to look somewhat worn after a few decades!

Total and spontaneous skin renewal will probably have to await the arrival of fully capable cell-repair nanomachines. These would be able to analyze whatever ails a particular cell or tissue component and restore it to its pristine condition. Along the road to that ideal, however, there will surely be jobs that more modest nanomachines can fill.

One of these will probably be the removal of wrinkles. *Wrinkles* arise from many causes, and it is unlikely that any single remedy will cure them all. It should be feasible to design special nanomachines to attack specific symptoms, however, and it would certainly be possible to employ a number of different kinds of nanomachines simultaneously.

Most wrinkles occur in response to *persistent folding*, as with crows feet around the eyes and smile lines, and the deep nasolabial folds that develop on either side of the mouth. Part of the problem is that the skin has lost elasticity, although there may be any number of contributing factors.

A start at treating one of the basic symptoms would be a nanomachine designed to improve the ski's elasticity, or at least to prevent further deterioration. What gives youthful skin the ability to adapt gracefully to the body's contours and movements while remaining tough enough to resist the wear and tar of everyday life is its supply of collagen fibers. Old skin is softer and floppier than young skin; it has less collagen, either because its cells make less collagen than they used to or because they produce too much collagenase, an enzyme that destroys collagen fibers. Most likely, both processes are at work.

To attack this problem, a nanomachine could be designed to seek out the cells that produce collagenase and slow or stop them. Another and possibly simpler approach would be a nanomachine that attaches itself to collagenase in such a way as to jam its collagen-cutting mechanism. Either way, it would be necessary to ensure that the nanomachines would act only in the skin, where we want them to, and to monitor the process carefully to keep it under control.

One simple way to keep the nanomachines from attacking *collagenase* anywhere but in the skin would be to make them temperature sensitive. They could remain inert and circulating in the bloodstream until they encountered a high temperature, say 110°F (easily obtained locally at the surface with a hot pack). Then they would migrate out of whatever capillary they were in and go to work.

As with the previous products, a *feature* of such a nanomachine would be its inability to reproduce. This would permit control of the action by limiting the dose. But whereas for other applications the degree of change was a matter of vanity, in this case it would be safeguard against removing so much collagenase as to make the skin leathery. Nonreproducing nanomachines also permit an incremental approach to the treatment. Instead of trying to prescribe the exact dose needed, the therapist could schedule a series of session in which no single application would contain enough of the product to do any harm.

Slender Now

A final example deals with what appears to be the number one beauty concern of American women: obesity. Here we venture beyond mere surface appearance and into bodily restructuring. And it must be admitted at the outset that not all kinds of obesity will yield to nanosurgical control. It appears that some obesity is genetic, some psychological, and some imaginary.

That said, however, there is still a vast market for a treatment that would short-circuit the ceaseless roller-coaster ride of conventional diet-splurge cycles. The ideal weight control product would allow a person to eat more or less what one wanted, continue to live a sedentary life, and maintain a figure that reasonably approximates one's ideal proportions. It should also be capable of spot application, to reduce a person's hips or abdomen for example, or wherever the problem appears to be.

At the same time, the product should induce a feeling of well-being rewarding the user with positive sensations to make it easy to stick with the program and keep using the product. Just how that might be accomplished I leave as an exercise for the reader. As a hint, consider the role of endorphins, those natural mood-altering chemicals that your body is constantly ready to produce.

One possible approach would enlist the aid of the body's own immune system. A nanomachine could be designed to home in on a fat cell, plant a marker on its surface that falsely identifies it as a pathogen, and let the nearest phagocyte take over. An easy way to accomplish this would be to instruct the nanomachine to fabricate the marker on the spot by altering a surface protein so that the cell is no longer recognizable as "part of the family."

To make sure that such a product works only where needed (a person needs some fat storage capability, after all), it would probably be wise to make this nanomachine temperature sensitive too. Surplus fat generally lies just under the skin, and it is notoriously good insulator, so it is easy to produce local temperature gradients across it. Any surplus nanomachines could just keep circulating until they were either eliminated naturally or put to use at some new problem elsewhere.

Techniques and Strategies

Any protein product designed to circulate through the body and perform some specific task on a certain class of cells must fulfill several criteria. It must first of all conform to a rigid set of specifications imposed by the body's immune system. It must also have some way of recognizing the target cells and, in some cases, some way of "knowing" when it has arrived at an appropriate location in the body.

Foiling the Immune System

The need to get past the immune system's detectors is a daunting requirement in view of that system's perpetual vigilance and almost devilish versatility. As we have learned more about the system's functioning, however, small chinks have begun to appear in its armor. Presumably these chinks will widen with further research, of which we can expect a considerable amount in the ten to twenty years that are likely to elapse before the first nanosurgical products appear.

However, there is one surefire way to protect such a product from the immune system, even in our present state of ignorance: package it is something that your immune system instantly recognizes as completely familiar. And what could be more familiar than red blood corpuscles of your own blood type? After all, routine blood transfusions trigger no immune response.

A case might be made for using white blood cells, since these are naturally able to leave the bloodstream, crawl through tissue, and sense their environment. It might appear that hitchhiking nanomachines could simply wait, reading the macrophage's sensory input until it finds itself at its intended destination, and then slip out to do its work.

There are at least three major drawbacks to this scheme. In the first place, blood contains about 600 red cells for every white cell. Getting enough white cells to do the job would involve either taking and separating a large volume of blood or cultivating large numbers of white cells outside the body. Second, it seems quite unlikely that information about the white blood cells' environment travels to some central organ analogous to a brain, where a hitchhiking nanomachine could read it. More than likely, the cell's sensory equipment is confined to the tips of its probing tendrils and is designed to be simply oblivious to contact with "self" materials of whatever kind. In other words it reacts only when it senses something it recognizes as foreign, and then it would react in exactly the same way, wherever it happens to be in the body. Finally, it may be possible to slip a nanomachine past a macrophage's defenses and into the cell at the outset, say by inactivating the white cell with cold. It is hard, however, to imagine how an emerging nanomachine could escape instant destruction by the fully active macrophage it had been ridding. It would be much like the case of the smiling young woman from Niger, who went for a ride on a tiger.

On the other hand, slipping nanomachines into red blood cells would be easy and completely harmless, assuming the product is correctly designed to be nontoxic and to penetrate the cell membrane without damaging the cell, as viruses do now. A sample of blood of the correct blood type would be separated into its components by centrifuge or otherwise. With no white cells to interfere, the nanomachines could infiltrate the red blood cells upon contact. A rinse to remove stragglers that failed to make contact would probably be in order before injection.

Packaging the nanomachines inside red blood cells gets them past the immune system, but it isolates them from many of the clues that might be used to activate them. Any attempt to poke a molecular fragment complex enough to act as a sensor out through the cell wall would invite attack by the immune system. What to do?

One possible approach would be to take advantage of the red blood cell's internal environment. All of the nanomachines we have considered are supposed to do their work at or near the body's surface. And that is where blood temperature is lowest and where the hemoglobin in red blood cell's becomes most oxygen-depleted. The nanomachines should be able to recognize these environmental clues while still inside the red blood cells.

Once the nanomachines get the "jump signal" that tells them they are in their "*drop zone*," they emerge from the red blood cell and begin to search for their programmed target cell, which they can recognize by its surface proteins. Some nanomachines may fall prey to wandering phagocytes, but most of them, having been released close to their targets, should succeed. Those that do will be safe from phagocytes as soon as they slip inside their target cells.

Sensory Equipment

So far we have postulated nanomachine molecules able to act on environmental cues such as temperature, oxygen depletion, and chemical concentration. The question is, how do we make molecular machines with these sensitivities? The answer is, with sensing mechanisms much like those found in natural molecules.

By now we have discovered many examples of proteins (enzymes) that bind to a certain shape of molecule, often with exquisite selectivity. The common feature of all these examples is a receptor site on one molecule that closely matches a shape on the other molecule. In most such instances, once the two molecules have come together they tend to stick. In other words, it takes only one molecule to fill the site and trigger a given reaction.

To make a sensor that responds only above a certain concentration of the target chemical, we need something more. We need to give the nanomachine two weakly binding receptors linked in such a way that both must be occupied before the nanomachine responds. That way nothing happens until the concentration is high enough to provide a reasonable chance that the second site will be filled before the molecule occupying the first one shakes loose. It would be possible to control the concentration required to trigger the device by adjusting the binding energy, and thereby the residence time, in the receptor sites.

For a sensor to detect above-normal temperatures, we need something similar but opposite. In this case we start with the active site already filled with a target molecule in such a way as to jam the nanomachine's mechanism. However, we design the attraction between the two so that they break apart under molecular bombardment above a certain temperature, freeing the nanomachine.

For a below-normal temperature sensor, we start with a nanomachine consisting of target and receptor molecules already connected by a flexible hinge, but with enough mismatch between their shapes that they are only weakly attracted at normal temperatures. As the temperature decreases, the two halves of this hinged ensemble spend more and more time in contact. Eventually, they stay together long enough for a slow-acting latch to function, holding them together permanently.

When a predetermined sequence of triggers fires, the nanomachine is activated to perform its intended function. We can envision the activated nanomachine unfolding itself into the required shape like one of those "transformer" toys, an obedient robot ready to go into action.

Intercommunications

For most of these applications, each nanomachine would be on its own, carrying out its instructions on atoms and molecules within reach of its manipulator arm or within a single target cell. For control over hair waviness (which requires that the nanomachines affect the hair's cross-sectional shape), however, hundreds of nanomachines in a particular follicle must work in concert. This means that they will have to communicate, at least in a rudimentary way.

Since such traits as hair colour and hair waviness are clearly inherited, it seems reasonable that the eventual channel for this communication would be genetic—by manipulating a person's DNA. But long before we have nanomachines capable of such fine control, we should be able to find ways to coordinate less-sophisticated groups of them to override the genetic instructions.

One such crude signaling method for hair waviness could be by means of a chemical gradient. The first nanomachine to gain a foot-hold in a particular follicle would stimulate its cell to generate a chemical marker, thereby staking its claim as the controlling nanomachine. This chemical marker would inhibit each subsequent nanomachine from emitting the same signal, thereby establishing a chemical gradient from one side of the hair follicle to the other. From this, each nanomachine could estimate its position with respect to the "master" nanomachine and modify its cell's hair-making activity accordingly.

Licensing

It should be clear that none of the techniques so far described, with the possible exception of the depilatory and tooth-protector nanomachines, could in any way be marketed over the counter. Each of the injectable products involves at least some technical sophistication and probably would need to be administrated by medical professionals. What will be the legal ramifications?

In the first place, as injectable pharmaceuticals, each of these products would come under the licensing authority of the Food and Drug Administration. (I think we an assume the continued jurisdiction of the FDA. This agency has, on the whole, done yeoman's service in protecting the American public from hazardous nostrums, despite recent criticism of its slowness in approving untried methods of controlling AIDS. And even totally useless government agencies have amazing powers of self-preservation.)

FDA approval is a long costly process, involving extensive testing for both safety and efficacy. This process would be undertaken primarily by drug companies, not cosmetic manufacturers, since externally applied cosmetics need to be certified only for safety.

At present, the preliminary testing of any such new technology would be done on animals, although there are many opposed to this practice. And there are valid arguments against animal tests in this case. It would not be easy to prove, for example, that a nanomachine designed to work in mice or rabbits would necessarily perform properly in humans.

At the same time, the idea of testing a brand new technology on humans is far from attractive and involves serious ethical questions. A possible way out may be to develop surrogate human tissues on which to perform these preliminary tests. Tissue culture methods so far developed, however, work only on single cell types. They are incapable, without considerable development, of producing the complex structures of even such a relatively simple organ as skin.

We would have no problem, of course, if we had self-reproducing cell-repair nanomachines to work with. They could monitor the internal workings of each cell in a growing population, reproducing in synchrony with the cell they inhabit, and guide the development of its descendants along predetermined pathways to produce any desired tissue component. Just for the testing of cosmetics alone, such technology could grow acres of human skin, complete with nerve endings, capillaries, hair follicles, sweat glands, and even subcutaneous musculature and fat cells—thriving under glass in racks of oversized Petri dishes in a laboratory. (It would also be important to cultivate mucous membranes, since these are much more sensitive to many irritants than the outer skin.) Incidentally, this system could also be used to grow all sorts of replacement organs for transplantation, relieving the chronic shortage. But by the time such capabilities matured, organ transplants are likely to be unnecessary except for the treatment of severe, life-threatening injuries. In-body repair, using injected nanosurgical nanomachines, would be far preferable in every way whenever there was a time for it to work, and it could be used to treat conditions we now tolerate until they become emergencies.

Long before cell-repair nanomachines become available however, a much simpler solution will have presented itself. Recent experiments show great progress in grafting human tissue onto animals without provoking an immune reaction. Carried just a bit further, this xenograft technique should enable us to produce living surrogate organisms whose tissues mimic those of humans closely enough for reliable test purposes but with just enough of a nervous system to control such vital function as respiration, digestion and circulation. Naturally, they would have no sensory equipment that would allow them to feel pain. It's hard to see how animal-right activists could object to laboratory tests on living nonanimals such as a pillow-shaped object covered with human skin.

ECONOMICS

What will be the likely response of the cosmetic industry? Probably not much—at first. As usual, the first feeble attempts will be greeted with derision and ascribed to the work of crackpots and charlatans. As the procedures became more mature and routine, however, and more affordable, the established companies might be forced to choose between opposing the new technology or buying in to it.

Some of the largest of them will probably hedge their bets by doing both. After all, Chevrolet completes as fiercely against Pontiac as it does against Ford. It will be interesting to watch the media battles that will predictably ensue.

DISTANT PROSPECTS

I have tried so far to touch on only the easily foreseeable and immediate consequences of the simplest nanosurgical nanomachines, but it is also worthwhile to sketch more distant possibilities. Once fully capable cell-repair nanomachines hit the market, there will be little the imagination can conjure up that would be impossible. Permanent rejuvenation of face and body is only the most obvious result.

Nonsurgical Nanomachine-based Cosmetics

In the end, however, it is unlikely that cosmetic nanosurgery will completely supplant conventional cosmetics. After all, there is something to be said for being able to adjust one's makeup to one's mood or costume, and to change back and forth between different looks quickly. And it should be noted that the primary consumers of beauty preparations today are not the elderly, who might presumably need them most, but the already young and beautiful.

Instead, it is likely that nanomachine techniques will take over the task of manufacturing cosmetic, a job for which programmable, self-reproducing nanomachines are ideally suited. The ability to manipulate individual atoms and promote or inhibit specific reactions could lead to all sorts of new fragrances and materials tailored to the cosmetic industry's needs.

Some cosmetics, in fact, might consist of nanomachines. Who could resist a fingernail polish or eye shadow, for example, made of units that automatically aligned and spaced themselves to produce a diffraction grating? Rainbow colours would flash from such a product, depending on the lighting and on the angle from which it was viewed. The colours might also be made to change with skin temperature or other environmental cues.

Total Makeover

Eventually, under the care of a competent practitioner backed up by talented molecular programmers and a responsive pharmaceutical industry, it should become possible to mold the face and body to whatever shape might be desired. Each person who cared to could achieve his or her own ideal of physical perfection or, for that matter, whatever frightening or gruesome effect they wanted.

Inevitably, there will be people who don't know how to leave well enough alone. Many who never liked their own youthful appearance will opt instead to copy some popular model or other sex symbol. It could become very confusing, with dozens of pop-idol look-alikes crowding the parks and boulevards of our future metropolis. Some may not relish the prospect, but we may never see the last of the Elvis clones.

11

ENGINES OF CONSTRUCTION

Protein engineering represents the first major step toward a more general capability for molecular engineering which would allow us to structure matter atom by atom. Coal and diamonds, sand and computer chips, cancer and healthy tissue: throughout history, variations in the arrangement of atoms have distinguished the cheap from the cherished, the diseased from the healthy. Arranged one way, atoms make up soil, air, and water; arranged another, they make up ripe strawberries. Arranged one way, they make up homes and fresh air; arranged another, they make up ash and smoke.

Our ability to arrange atoms lies at the foundation of technology. We have come far in our atom arranging, from chipping flint for arrowheads to machining aluminum for spaceships. We take pride in our technology, with our lifesaving drugs and desktop computers. Yet our spacecraft are still crude, our computers are still stupid, and the molecules in our tissues still slide into disorder, first destroying health, then life itself. For all our advances in arranging atoms, we still use primitive methods. With our present technology, we are still forced to handle atoms in unruly herds. But the laws of nature leave plenty of room for progress, and the pressures of world competition are even now pushing us forward. For better or for worse, the greatest technological breakthrough in history is still to come.

TWO STYLES OF TECHNOLOGY

Our modern technology builds on an ancient tradition. Thirty thousand years ago, chipping flint was the high technology of the day. Our ancestors grasped stones containing trillions of trillions of atoms and removed chips containing billions of trillions of atoms to make their axheads; they made fine work with skills difficult to imitate today. They also made patterns on cave walls in France with sprayed paint, using their hands as stencils. Later they made pots by baking clay, then bronze by cooking rocks. They shaped bronze by pounding it. They made iron, then steel, and shaped it by heating, pounding, and removing chips.

We now cook up pure ceramics and stronger steels, but we still shape them by pounding, chipping, and so forth. We cook up pure silicon, saw it into slices, and make patterns on its surface using tiny stencils and sprays of light. We call the products "chips" and we consider them exquisitely small, at least in comparison to axheads. Our microelectronic technology has managed to stuff machines as powerful as the room-sized computers of the early 1950 onto a few silicon chips in a pocket-sized computer. Engineers are now making ever smaller devices, slinging herds of atoms at a crystal surface to build up wires and components one tenth the width of a fine hair.

These microcircuits may be small by the standards of flint chippers, but each transistor still holds trillions of atoms, and so-called "microcomputers" are still visible to the naked eye. By the standards of a newer, more powerful technology they will seem gargantuan.

The ancient style of technology that led from flint chips to silicon chips handles atoms and molecules in bulk; call it *bulk technology*. The new technology will handle individual atoms and molecules with control and precision; call it *molecular technology*. It will change our world in more ways than we can imagine.

Microcircuits have parts measured in mice meters—that is, in millionths of a meter—but molecules are measured in nanometers (a thousand times smaller). We can use the terms "nanotechnology" and "molecular technology" interchangeably to describe the new style of technology. The engineers of the new technology will build both nano circuits and nanomachines.

Molecular Technology Today

One dictionary definition of a *machine* is "any system, usually of rigid bodies, formed and connected to alter, transmit, and direct applied forces in a predetermined manner to accomplish a specific objective, such as the performance of useful work." *Molecular machines* fit this definition quite well.

To imagine these machines, one must first picture molecules. We can picture atoms as beads and molecules as clumps of beads, like a child's beads linked by snaps. In fact, chemists do sometimes visualize molecules by building models from plastic beads (some of which link in several directions, like the hubs in a Tinkertoy set). Atoms are rounded like beads, and although molecular bonds are not snaps, our picture at least captures the essential notion that bonds can be broken and reformed.

If an atom were the size of a small marble, a fairly complex molecule would be the size of your fist. This makes a useful mental image, but atoms are really about 1/10,000 the size of bacteria, and bacteria are about 1/10,000 the size of mosquitoes. (An atomic nucleus, however, is about 1/100,000 the size of (the atom itself; the difference between an atom and its nucleus is the difference between a fire and a nuclear reaction.)

The things around us act as they do because of the way their molecules behave. Air holds neither its shape nor its volume because its molecules move freely, bumping and ricocheting through open space. Water molecules stick together as they move about, so water holds a constant volume as it changes shape. Copper holds its shape because its atoms stick together in regular patterns; we can bend it and hammer it because its atoms can slip over one another while remaining bound together. Glass shatters when we hammer it be cause its atoms separate before they slip. Rubber consists of networks of kinked molecules, like a tangle of springs. When stretched and released, its molecules straighten and then coil again. These simple molecular patterns make up passive substances. More complex patterns make up the active nanomachines of living cells.

Biochemists already work with these machines, which are chiefly made of protein, the main engineering material of living cells. These molecular machines have relatively few atoms, and so they have lumpy surfaces, like objects made by gluing together a handful of small marbles. Also, many pairs of atoms are linked by bonds that can bend or rotate, and so protein machines are unusually flexible. But like all machines, they have parts of different shapes and sizes that do useful work. All machines use clumps of atoms as parts. Protein machines simply use very small clumps.

Biochemists dream of designing and building such devices, but there are difficulties to be overcome. Engineers use beams of light to project patterns onto silicon chips, but chemists must build much more indirectly than that. When they combine molecules in various sequences, they have only limited control over how the molecules join. When biochemists need complex molecular machines, they still have to borrow them from cells. Nevertheless, advanced molecular machines will eventually let them build nanocircuits and nanomachines as easily and directly as engineers now build microcircuits or washing machines. Then progress will become swift and dramatic.

Genetic engineers are already showing the way. Ordinarily, when chemists make molecular chains—called "*polymers*"—they dump molecules into a vessel where they bump and snap together haphazardly in a liquid. The resulting chains have varying lengths, and the molecules are strung together in no particular order.

But in modem gene synthesis machines, genetic engineers build more orderly polymers—specific DNA molecules—by combining molecules in a particular order. These molecules are the nucleotides of DNA (the letters of the genetic alphabet) and genetic engineers don't dump them all in together. Instead, they direct the machine to add different nucleotides in a particular sequence to spell out a particular message. They first bond one kind of nucleotide to the chain ends, then wash away the leftover material and add chemicals to prepare the chain ends to bond the next nucleotide. They grow chains as they bond on nucleotides, one at a time, in a programmed sequence. They anchor the very first nucleotide in each chain to a solid surface to keep the chain from washing away with its chemical bathwater. In this way, they have a big clumsy machine in a cabinet assemble specific molecular structures from parts a hundred million times smaller than itself.

But this blind assembly process accidentally omits nucleotides from some chains. The likelihood of mistakes grows as chains grow longer. Like workers discarding bad parts before assembling a car, genetic engineers reduce errors by discarding bad chains. Then, to join these short chains into working genes (typically thousands of nucleotides long), they turn to molecular machines found in bacteria.

These protein machines, called *restriction enzymes*, "read" certain DNA sequences as "cut here." They read these genetic patterns by touch, by sticking to them, and they cut the chain by rearranging a few atoms. Other enzymes splice pieces together, reading matching parts as "glue here"—likewise "reading" chains by selective stickiness and splicing chains by rearranging a few atoms. By using gene machines to write, and restriction enzymes to cut and paste, genetic engineers can write and edit whatever DNA messages they choose.

But by itself, DNA is a fairly worthless molecule. It is neither strong like Kevlar, nor colourful like a dye, nor active like an enzyme,

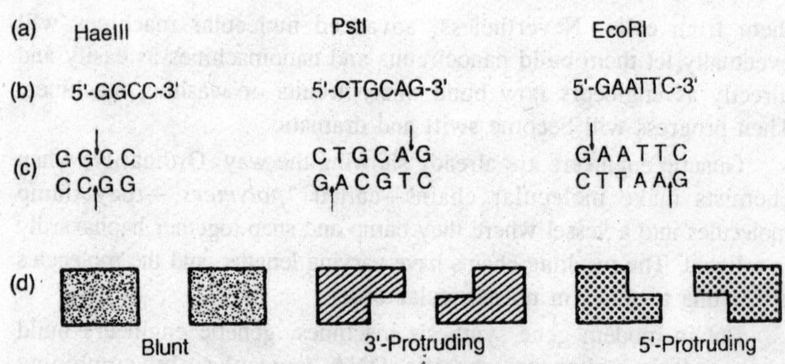

Fig. 11.1 Restriction enzymes.

yet it has something that industry is prepared to spend millions of dollars to use: the ability to direct molecular machines called *ribosomes*. In cells, molecular machines first transcribe DNA, copying its information to make RNA "tapes." Then, much as old numerically controlled machines shape metal based on instructions stored on tape, ribosomes build proteins based on instructions stored on RNA strands. And proteins are useful.

Proteins, like DNA, resemble strings of lumpy beads. But unlike DNA, protein molecules fold up to form small objects able to do things. Some are enzymes, machines that build up and tear down molecules (and copy DNA, transcribe it, and build other proteins in the cycle of life). Other proteins are hormones, binding to yet other proteins to signal cells to change their behaviour. Genetic engineers can produce these objects cheaply by directing the cheap and efficient molecular machinery inside living organisms to do the work. Whereas engineers running a chemical plant must work with vats of reacting chemicals (which often misarrange atoms and make noxious by-products), engineers working with bacteria can make them absorb chemicals, carefully rearrange the atoms, and store a product or re lease it into the fluid around them.

Genetic engineers have now programmed bacteria to make proteins ranging from human growth hormone to rennin, an enzyme used in making cheese. The pharmaceutical company Eli Lilly (Indianapolis) is now marketing *Humulin*, *human insulin* molecules made by bacteria.

Existing Protein Machines

These protein hormones and enzymes selectively stick to other molecules. An enzyme changes its target's structure, then moves on; a hormone affects its target's behaviour only so long as both remain stuck together. Enzymes and hormones can be described in mechanical terms, but their behaviour is more often described in chemical terms.

But other proteins serve basic mechanical functions. Some push and pull, some act as cords or struts, and parts of some molecules make excellent bearings. The machinery of muscle, for instance, has gangs of proteins that reach, grab a "rope" (also made of protein), pull it, then reach out again for a fresh grip; whenever you move, you use these machines. Amoebas and human cells move and change shape by using fibers and rods that act as molecular muscles and bones. A reversible, variable-speed motor drives bacteria through water by turning a corkscrew-shaped propeller. If a hobbyist could build tiny cars around such motors, several billions of billions would fit in a pocket, and 150-lane freeways could be built through your finest capillaries.

Simple molecular devices combine to form systems resembling industrial machines. In the 1950s engineers developed machine tools that cut metal under the control of a punched paper tape. A century and a half earlier, Joseph-Marie Jacquard had built a loom that wove complex patterns under the control of a chain of punched cards. Yet over three billion years before Jacquard, cells had developed the machinery of the ribosome. Ribosomes are proof that *nanomachines* built of protein and RNA can be programmed to build complex molecules.

Then consider *viruses*. One kind, the T4 phage, acts like a spring-loaded syringe and looks like something out of an industrial parts catalog. It can stick to a bacterium, punch a hole, and inject viral DNA (yes, even bacteria suffer infections). Like a conqueror seizing factories to build more tanks, this DNA then directs the cell's machines to build more viral DNA and syringes. Like all organisms, these viruses exist because they are fairly stable and are good at getting copies of themselves made.

Whether in cells or not, nanomachines obey the universal laws of nature. Ordinary chemical bonds hold their atoms together, and ordinary chemical reactions (guided by other nanomachines) assemble them. Protein molecules can even join to form machines without special help, driven only by thermal agitation and chemical forces. By mixing viral proteins (and the DNA they serve) in a test tube, molecular biologists have assembled working T4 viruses. This ability is surprising: imagine putting automotive parts in a large box, shaking it, and finding an assembled car when you look inside! Yet the T4 virus is but one of many self-assembling structures. Molecular biologists have taken the machinery of the ribosome apart into over fifty separate protein and RNA molecules, and then combined them in test tubes to form working ribosomes again.

To see how this happens, imagine different T4 protein chains floating around in water. Each kind folds up to form a lump with distinctive bumps and hollows, covered by distinctive patterns of oiliness, wetness, and electric charge. Picture them wandering and tumbling, jostled by the thermal vibrations of the surrounding water molecules. From time to time two bounce together, then bounce apart. Some times, though, two bounce together and fit, bumps in hollows, with sticky patches matching; they then pull together and stick. In this way protein adds to protein to make sections of the virus, and sections assemble to form the whole.

Protein engineers will not need *nanoarms* and *nanohands* to assemble complex *nanomachines*. Still, tiny manipulators will be useful and they will be built. Just as today's engineers build machinery as complex as player pianos and robot arms from ordinary motors, bearings, and moving parts, so tomorrow's biochemists will be able to use protein molecules as motors, bearings, and moving parts to build robot arms which will themselves be able to handle individual molecules.

Designing with Protein

How far off is such an ability? Steps have been taken, but much work remains to be done. Biochemists have already mapped the structures of many proteins. With gene machines to help write DNA tapes, they can direct cells to build any protein they can design. But they still do not know how to design chains that will fold up to make proteins of the right shape and function. The forces that fold proteins are weak, and the number of plausible ways a protein might fold is astronomical, so designing a large protein from scratch is not easy.

The forces that stick proteins together to form complex machines are the same ones that fold the protein chains in the first place. The differing shapes and kinds of stickiness of amino acids—the lumpy molecular "*beads*" forming protein chains—make each protein chain fold up in a specific way to form an object of a particular shape. Biochemists have learned rules that suggest how an amino acid chain might fold, but the rules are not very firm. Trying to predict how a chain will fold is like trying to work a jigsaw puzzle, but a puzzle with no pattern printed on its pieces to show when the fit is correct, and with pieces that seem to fit together about as well (or as badly) in many different ways, all but one of them wrong. False starts could consume many lifetimes, and a correct answer might not even be recognized. Biochemists using the best computer programs now available still cannot predict how a long, natural protein chain will actually fold, and some of them have despaired of designing protein molecules soon.

Yet most biochemists work as scientists, not as engineers. They work at predicting how natural proteins will fold, not at *designing* proteins that will fold predictably. These tasks may sound similar, but they differ greatly: the first is a scientific challenge, the second is an engineering challenge. Why should natural proteins fold in a way that scientists will find easy to predict? All that nature requires is that they in fact fold correctly, not that they fold in a way obvious to people.

Proteins could be designed from the start with the goal of making their folding more predictable. Carl Pabo, writing in the journal *Nature*, has suggested a design strategy based on this insight, and some biochemical engineers have designed and built short chains of a few dozen pieces that fold and nestle onto the surfaces of other molecules as planned. They have designed from scratch a protein with properties like those of melittin, a toxin in bee venom. They have modified existing enzymes, changing their behaviours in predict able ways. Our understanding of proteins is growing daily.

In 1959, according to biologist Garrett Hardin, some geneticists called genetic engineering impossible; today, it is an industry. Biochemistry and computer-aided design are now exploding fields, and as Frederick Blattner wrote in the journal *Science*, "computer chess programs have already reached the level below the grand master. Perhaps the solution to the protein-folding problem is nearer than we think." William Rastetter of Genentech, writing in *Applied Biochemistry* and *Biotechnology*, asks, "How far off is *de novo* enzyme design and synthesis? Ten, fifteen years?" He answers, "Perhaps not that long."

Forrest Carter of the U.S. Naval Research Laboratory, An Aviram and Philip Seiden of IBM, Kevin Ulmer of Genex Corporation, and other researchers in university and industrial laboratories around the globe have already begun theoretical work and experiments aimed at developing molecular switches, memory devices, and other structures that could be incorporated into a protein-based computer. The U.S. Naval Research Laboratory has held two international work shops on molecular electronic devices, and a meeting sponsored by the U.S.

National Science Foundation has recommended support for basic research aimed at developing molecular computers. Japan has reportedly begun a multimillion-dollar program aimed at developing self-assembling molecular motors and computers, and VLSI Research Inc., of San Jose, reports that "It looks like the race to bio-chips [term for molecular electronic systems] has already started. NEC, Hitachi, Toshiba, Matsushita, Fujitsu, Sanyo-Denki and Sharp have commenced full-scale research efforts on bio-chips for bio-computers."

Biochemists have other reasons to want to learn the art of protein design. New enzymes promise to perform dirty, expensive chemical processes more cheaply and cleanly, and novel proteins will offer a whole new spectrum of tools to biotechnologists. We are already on the road to protein engineering, and as Kevin Ulmer notes in the quote from Science that heads this chapter, this road leads "toward a more general capability for molecular engineering which would allow us to structure matter atom by atom."

Second-Generation Nanotechnology

Despite its versatility, protein has shortcomings as an engineering material. Protein machines quit when dried, freeze when chilled, and cook when heated. We do not build machines of flesh, hair, and gelatin; over the centuries, we have learned to use our hands of flesh and bone to build machines of wood, ceramic, steel, and plastic. We will do likewise in the future. We will use protein machines to build nano machines of tougher stuff than protein.

As nanotechnology moves beyond reliance on proteins, it will grow more ordinary from an engineer's point of view. Molecules will be assembled like the components of an erector set, and well- bonded parts wili stay put. Just as ordinary tools can build ordinary machines from parts, so molecular tools will bond molecules together to make tiny gears, motors, levers, and casings, and assemble them to make complex machines.

Parts containing only a few atoms will be lumpy, but engineers can work with lumpy parts if they have smooth bearings to support them. Conveniently enough, some bonds between atoms make fine bearings; a part can be mounted by means of a single chemical bond that will let it turn freely and smoothly. Since a bearing can be made using only two atoms (and since moving parts need have only a few atoms), nanomachines can indeed have mechanical components of molecular size.

How will these better machines be built? Over the years, engineers have used technology to improve technology. They have used metal tools to shape metal into better tools, and computers to design and program better computers. They will likewise use protein nano machines to build better nanomachines. Enzymes show the way: they assemble large molecules by "grabbing" small molecules from the water around them, then holding them together so that a bond forms. Enzymes assemble DNA, RNA, proteins, fats, hormones, and chlorophyll in this way—indeed, virtually the whole range of molecules found in living things.

Biochemical engineers, then, will construct new enzymes to assemble new patterns of atoms. For example, they might make an enzyme- like machine which will add carbon atoms to a small spot, layer on layer. If bonded correctly, the atoms will build up to form a fine, flexible diamond fiber having over fifty times as much strength as the same weight of aluminum. Aerospace companies will line up to buy such fibers by the ton to make advanced composites. (This shows one small reason why military competition will drive molecular technology forward, as it has driven so many fields in the past.)

But the great advance will come when protein machines are able to make structures more complex than mere fibers. These programmable protein machines will resemble ribosomes programmed by RNA, or the older generation of automated machine tools programmed by punched tapes. They will open a new world of possibilities, letting engineers escape the limitations of proteins to build rugged, compact machines with straightforward designs. Engineered proteins will split and join molecules as enzymes do. Existing proteins bind a variety of smaller molecules, using them as chemical tools; newly engineered proteins will use all these tools and more.

Further, organic chemists have shown that chemical reactions can produce remarkable results even without nanomachines to guide the molecules. Chemists have no direct control over the tumbling motions of molecules in a liquid, and so the molecules are free to react in any way they can, depending on how they bump together. Yet chemists nonetheless coax reacting molecules to form regular structures such as cubic and dodecahedral molecules, and to form unlikely-seeming structures such as molecular rings with highly strained bonds. Molecular machines will have still greater versatility in bond making, because they can use similar molecular motions to make bonds, but can guide these motions in ways that chemists cannot.

Indeed, because chemists cannot yet direct molecular motions, they can seldom assemble complex molecules according to specific plans. The largest molecules they can make with specific, complex patterns are all linear chains. Chemists form these patterns (as in gene machines) by adding molecules in sequence, one at a time, to a growing chain. With only one possible bonding site per chain, they can be sure to add the next piece in the right place.

But if a rounded, lumpy molecule has (say) a hundred hydrogen atoms on its surface, how can chemists split off just one *particular* atom (the one five up and three across from the bump on the front) to add something in its place? Stirring simple chemicals together will seldom do the job, because small molecules can seldom select specific places to react with a large molecule. But protein machines will be more choosy.

A flexible, programmable protein machine will grasp a large molecule (the work piece) while bringing a small molecule up against it in just the right place. Like an enzyme, it will then bond the molecules together. By bonding molecule after molecule to the workpiece, the machine will assemble a larger and larger structure while keeping complete control of how its atoms are arranged. This is the key ability that chemists have lacked.

Like ribosomes, such nanomachines can work under the direction of molecular tapes. Unlike ribosomes, they will handle a wide variety of small molecules (not just amino acids) and will join them to the work piece anywhere desired, not just to the end of a chain. Protein machines will thus combine the splitting and joining abilities of enzymes with the programmability of ribosomes. But whereas ribosomes can build cnly the loose folds of a protein, these protein machines will build small, solid objects of metal, ceramic, or diamond—invisibly small, but rugged.

Where our fingers of flesh are likely to bruise or burn, we turn to steel tongs. Where protein machines are likely to crush or disintegrate, we will turn to nanomachines made of tougher stuff.

Universal Assemblers

These second-generation nanomachines—built of more than just proteins-—will do all that proteins can do, and more. In particular, some will serve as improved devices for assembling molecular structures. Able to tolerate acid or vacuum, freezing or baking, depending on design, enzyme-like second-generation machines will be able to use as "tools" almost any of the reactive molecules used by chemists—but they will wield them with the precision of programmed machines. They will be able to bond atoms together in virtually any stable pattern, adding a few at a time to the surface of a work piece until a complex structure is complete. Think of such nanomachines as *assemblers*.

Because assemblers will let us place atoms in almost any reason able arrangement, they will let us build almost anything that the laws of nature allow to exist. In particular, they will let us build almost anything we can design—including more assemblers. The consequences of this will be profound, because our crude tools have let us explore only a small part of the range of possibilities that natural law permits. Assemblers will open a world of new technologies.

Advances in the technologies of medicine, space, computation, and production—and warfare—all depend on our ability to arrange atoms. With assemblers, we will be able to remake our world or destroy it. So at this point it seems wise to step back and look at the prospect as clearly as we can, so we can be sure that assemblers and nanotechnology are not a mere futurological mirage.

Nailing Down Conclusions

Principle of Quantum Physics in Molecular Machines

This principle states (among other things) that particles can't be pinned down in an exact location for any length of time. It limits what molecular machines can do, just as it limits what anything else can do. Nonetheless, calculations show that the uncertainty principle places few important limits on how well atoms can be held in place, at least for the purposes outlined here. The uncertainty principle makes *electron* positions quite fuzzy, and in fact this fuzziness deter mines the very size and structure of atoms. An atom as a whole, however, has a comparatively definite position set by its comparatively massive nucleus. If atoms didn't stay put fairly well, molecules would not exist. One needn't study quantum mechanics to trust these conclusions, because molecular machines in the cell demonstrate that molecular machines work.

Molecular Vibrations

Thermal vibrations will cause greater problems than will the uncertainty principle, yet here again existing molecular machines directly demonstrate that molecular machines can work at ordinary temperatures. Despite thermal vibrations, the *DNA-copying* machinery in some cells makes less than one error in 100,000,000,000 operations.

To achieve this accuracy, however, cells use machines (such as the enzyme DNA polymerase I) that proofread the copy and correct errors. Assemblers may well need similar error-checking and error-correcting abilities, if they are to produce reliable results.

Disruption of Molecular Machines

High-energy radiation can break chemical bonds and disrupt molecular machines. Living cells once again show that solutions exist: they operate for years by repairing and replacing radiation-damaged parts. Because individual machines are so tiny, however, they present small targets for radiation and are seldom hit. Still, if a system of nanomachines must be reliable, then it will have to tolerate a certain amount of damage, and damaged parts must regularly be repaired or replaced. This approach to reliability is well known to designers of aircraft and spacecraft.

Assemblers

The working of molecular machinery of cells makes a simple and powerful case that natural law permits small clusters of atoms to behave as controlled machines, able to build other nano machines. Yet de spite their basic resemblance to ribosomes, assemblers will differ from anything found in cells; the things they do—while consisting of ordinary molecular motions and reactions—will have novel results. No cell, for example, makes diamond fiber.

The idea that new kinds of nanomachinery will bring new, useful abilities may seem startling: in all its billions of years of evolution, life has never abandoned its basic reliance on protein machines. Does this suggest that improvements are impossible, though? Evolution progresses through small changes, and evolution of DNA cannot easily replace DNA. Since the DNA/RNA/ribosome system is specialized to make proteins, life has had no real opportunity to evolve an alternative. Any production manager can well appreciate the reasons; even more than a factory, life cannot afford to shut down to replace its old systems.

Improved molecular machinery should no more surprise us than alloy steel being ten times stronger than bone, or copper wires transmitting signals a million times faster than nerves. Cars out speed cheetahs, jets out fly falcons, and computers already out calculate head- scratching humans. The future will bring further examples of improvements on biological evolution, of which second-generation nanomachines will be but one.

In physical terms, it is clear enough why advanced assemblers will be able to do more than existing protein machines. They will be programmable like ribosomes, but they will be able to use a wider range of tools than all the enzymes in a cell put together. Because they will be made of materials far more strong, stiff, and stable than proteins, they will be able to exert greater forces, move with greater precision, and endure harsher conditions. Like an industrial robot arm—but unlike anything in a living cell—they will be able to rotate and move molecules in three dimensions under programmed control, making possible the precise assembly of complex objects. These advantages will enable them to assemble a far wider range of molecular structures than living cells have done.

One might doubt that *artificial nanomachines* could even equal the abilities of nano machines in the cell, if there were reason to think that cells contained some special magic that makes them work. This idea is called "*vitalism.*" Biologists have abandoned it because they have found chemical and physical explanations for every aspect of living cells yet studied, including their motion, growth, and reproduction. Indeed, this knowledge is the very foundation of biotechnology.

Nanomachines floating in sterile test tubes, free of cells, have been made to perform all the basic sorts of activities that they perform inside living cells. Starting with chemicals that can be made from smoggy air, biochemists have built working protein machines with out help from cells. R. B. Merrifield, for example, used chemical techniques to assemble simple amino acids to make bovine pancreatic ribonuclease, an enzymatic device that disassembles RNA molecules. Life is special in structure, in behaviour, and in what it feels like from the inside to be alive, yet the laws of nature that govern the machinery of life also govern the rest of the universe.

Sheer curiosity seems reason enough to examine the possibilities opened by nanotechnology, but there are stronger reasons. These developments will sweep the world within ten to fifty years—that is, within the expected lifetimes of ourselves or our families. What is more, the conclusions of the following chapters suggest that a wait-and-see policy would be very expensive—that it would cost many millions of lives, and perhaps end life on Earth.

Feasibility of Nanotechnology

Is the case for the feasibility of nanotechnology and assemblers firm enough that they should be taken seriously? It seems so, because the heart of the case rests on two well-established facts of science and engineering. These are (1) that existing molecular machines serve a range of basic functions, and (2) that parts serving these basic functions can be combined to build complex machines. Since chemical reactions can bond atoms together in diverse ways, and since molecular machines can direct chemical reactions according to programmed instructions, assemblers definitely are feasible.

Nanocomputers

Assemblers will bring one breakthrough of obvious and basic importance: engineers will use them to shrink the size and cost of computer circuits and speed their operation by enormous factors.

With today's bulk technology, engineers make patterns on silicon chips by throwing atoms and photons at them, but the patterns re main flat and molecular-scale flaws are unavoidable. With assemblers, however, engineers will build circuits in three dimensions, and build to atomic precision. The exact limits of electronic technology today remain uncertain because the quantum behaviour of electrons in complex networks of tiny structures presents complex problems, some of them resulting directly from the uncertainty principle. Whatever the limits are, though, they will be reached with the help of assemblers.

The fastest computers will use electronic effects, but the smallest may not. This may seem odd, yet the essence of computation has nothing to do with electronics. A digital computer is a collection of switches able to turn one another on and off. Its switches start in one pattern (perhaps representing 2 + 2), then switch one another into a new pattern (representing 4), and so on. Such patterns can represent almost anything. Engineers build computers from tiny electrical switches connected by wires simply because mechanical switches connected by rods or strings would be big, slow, unreliable, and expensive, today.

The idea of a purely mechanical computer is scarcely new. In England during the mid-1800s, Charles Babbage invented a mechanical computer built of brass gears; his co-worker Augusta Ada, the Countess of Lovelace, invented computer programming. Babbage's endless redesigning of the machine, problems with accurate manufacturing, and opposition from budget-watching critics (some doubting the usefulness of computers!), combined to prevent its completion.

In this tradition, Danny Hill is and Brian Silverman of the MIT Artificial Intelligence Laboratory built a special-purpose mechanical computer able to play tic-tac-toe. Yards on a side, full of rotating shafts and movable frames that represent the state of the board and the strategy of the game, it now stands in the Computer Museum in Boston. It looks much like a large ball-and-stick molecular model, for it is built of Tinkertoys.

Brass gears and Tinkertoys make for big, slow computers. With components a few atoms wide, though, a simple mechanical computer would fit within 1/100 of a cubic micron, many billions of times more compact than today's so-called microelectronics. Even with a billion bytes of storage, a nano mechanical computer could fit in a box a micron wide, about the size of a bacterium. And it would be fast Although mechanical signals move about 100,000 times slower than the electrical signals in today's machines, they will need to travel only 1/1,000,000 as far, and thus will face less delay. So a mere mechanical computer will work faster than the electronic whirl winds of today.

Electronic nanocomputers will likely be thousands of times faster than electronic microcomputers—perhaps hundreds of thousands of times faster, if a scheme proposed by Nobel Prize-winning physicist Richard Feynman works out. Increased speed through decreased size is an old story in electronics.

Disassemblers

Molecular computers will control molecular assemblers, providing the swift flow of instructions needed to direct the placement of vast numbers of atoms. Nanocomputers with molecular memory devices will also store data generated by a process that is the opposite of assembly.

Assemblers will help engineers synthesize things; their relatives, disassemblers, will help scientists and engineers analyze things. The case for assemblers rests on the ability of enzymes and chemical reactions to form bonds, and of machines to control the process. The case for disassemblers rests on the ability of enzymes and chemical reactions to break bonds, and of machines to control the process. Enzymes, acids, oxidizers, alkali metals, ions, and reactive groups of atoms called free radicals–all can break bonds and remove groups of atoms. Because nothing is absolutely immune to corrosion, it seems that molecular tools will be able to take anything apart, a few atoms at a time. What is more, a nanomachine could (at need or convenience) apply mechanical force as well, in effect prying groups of atoms free.

A nanomachine able to do this, while recording what it removes layer by layer, is a *disassembler*. Assemblers, disassemblers, and nano computers will work together. For example, a nano computer system will be able to direct the disassembly of an object, record its structure, and then direct the assembly of perfect copies. And this gives some hint of the power of nanotechnology.

World Made New

Assemblers will take years to emerge, but their emergence seems almost inevitable: Though the path to assemblers has many steps, each step will bring the next in reach, and each will bring immediate rewards. The first steps have already been taken, under the names of "genetic engineering" and "biotechnology." Other paths to assemblers seem possible. Barring worldwide destruction or worldwide controls, the technology race will continue whether we wish it or not. And as advances in computer-aided design speed the development of molecular tools, the advance toward assemblers will quicken.

To have any hope of understanding our future, we must under stand the consequences of assemblers, disassemblers, and nano-computers. They promise to bring changes as profound as the industrial revolution, antibiotics, and nuclear weapons all rolled up in one massive breakthrough. To understand a future of such profound change, it makes sense to seek principles of change that have survived the greatest upheavals of the past. They will prove a useful guide.

12

Computing with Molecules

How fast and powerful can computers become? Will it be possible someday to create artificial '*brains*" that have intellectual capabilities comparable—or even superior—to those of human beings? The answers to these questions depend to a very great extent on a single factor: how small and dense we can make computer circuits.

Few if any researchers believe that our present technology—*semiconductor-based solid-state microelectronics*—will lead to circuitry dense and complex enough to give rise to true cognitive abilities. And until recently, none of the technologies proposed as successors to solid-state microelectronics had shown enough promise to rise above the pack. Within the past year, however, scientists have achieved revolutionary advances that may very well radically change the future of computing. And although the road from here to intelligent machines is still rather long and might turn out to have unbridgeable gaps, the fact that there is a potential path at all is something of a triumph.

Microcircuits

The recent advances were in molecular-scale electronics, a field emerging around the premise that it is possible to build individual molecules that can perform functions identical or analogous to those of the transistors, diodes, conductors and other key components of today's microcircuit. After a period of high hopes but few tangible results, several developments over the past few years have raised expectations that this technology may one day provide the building blocks for future generations of ultrasmall, ultradense electronic computer logic. In a remarkable series of demonstrations, chemists, physicists and engineers have shown that individual molecules can conduct and switch electric current and store information.

Electronic Switches

In 1999, in an achievement widely reported in the popular press, researchers from Hewlett-Packard and the University of California at Los Angeles announced that they had built an *electronic switch* consisting of a layer of several million molecules of an organic substance called *rotaxane*. By linking a number of switches, the researchers produced a rudimentary version of an AND gate, a device that performs a basic logic operation. With well over a million molecules apiece, the switches are far larger than would be desirable. And they could be switched only one time before becoming inoperable. Nevertheless, their assembly into a logic gate was of fundamental significance.

Within months of that announcement, our groups at Yale and Rice universities published results on a different class of molecules that acted as a reversible switch. And one month later we described a molecule we had created that could change its electrical conductivity by storing electrons on demand, acting as a memory device.

To produce our switch, we inserted regions into the molecules that trapped electrons, but only when the molecules were subjected to certain voltages. Thus, the degree to which the molecules resisted a flow of electrons depended on the voltage applied to them. In fact, by varying the voltage, we could repeatedly change the molecules at will from a conducting to a nonconducting state—which is the basic requirement for an electrical switch. The tiny device actually consisted of a layer of about 1,000 molecules of nitroamine benzenethiol sandwiched between metal contacts.

Electron Sucker

After creating the switch, we realized that if we could redesign the molecule so that it could retain electrons rather than trapping them briefly, we would have something that could work as a memory element. We went to work on the trapping region of the molecule, modifying it so that its conductivity could be changed repeatedly. The resulting "*electron sucker*" could retain electrons for nearly 10 minutes—compared with a few milliseconds for conventional silicon-based dynamic random-access memory.

Although the advances were encouraging, the challenges remaining are enormous. Creating individual devices is an essential first step. But before we can build complete, useful circuits we must find a way to secure many millions, if not billions, of molecular devices of various types against some kind of immobile surface and to link them in any manner and into whatever patterns our circuit diagrams dictate. The technology is still too young to say for sure whether this monumental challenge will ever be surmounted.

Given the magnitude of the challenges ahead, why did researchers and even the mainstream media pay so much attention to the recent advances? The answer has to do with industrial society's dependence on microelectronics—and the limits of the form of the technology we have today.

TRANSISTORS

That form—solid-state and silicon-based—follows one of the most famous axioms in technology: Moore's Law. It relates that the number of transistors that can be fabricated on a silicon integrated circuit—and therefore the computing speed of such a circuit—is doubling every 18 to 24 months, After following this remarkable curve for four decades, solid-state microelectronics has advanced to the point at which engineers can now put on a sliver of silicon of just a few square centimeters some 100 million transistors, with key features measuring 0.18 micron.

These transistors are still far larger than molecular-scale devices. To put the size differential in perspective, if the conventional transistor were scaled up so that it occupied the printed page you are reading, a molecular device would be the period at the end of this sentence. Even in a dozen years, when industry projections suggest that silicon transistors will have shrunk to about 1.20 nanometers in length, they will still be more than 60,000 times larger in area than molecular electronic devices.

Moreover, no one expects conventional silicon-based micro electronics to continue following Moore's Law forever. At some point, chip-fabrication specialists will find it economically infeasible to continue scaling down microelectronics. As they pack more transistors onto a chip, phenomena such as stray signals on the chip, the need to dissipate the heat from so many closely packed devices, and the difficulty of creating the devices in the first place will halt or severely slow progress.

Indeed, various nagging (though not yet fundamental) problems in the fabrication of efficient smaller *Silicon transistors* and their interconnections are becoming increasingly bother some. Many experts expect these challenges to intensify dramatically as the transistors approach the 0.1-micron level. Because of these and other difficulties, the exponential increase in transistor densities and processing rates of integrated circuits is being sustained only by a similar exponential rise in the financial outlays necessary to build the facilities that produce these chips. Eventually the drive to down scale will run headlong into these extreme facility costs, and the market will reach equilibrium. Many experts project that this will happen around or before 2015, when a fabrication facility is projected to cost nearly $200 billion. When that happens, the long period of breathtaking advances in the processing power of computer chips will have run its course. Further increases in the power of the chips will be prohibitively costly Unfortunately, this impasse will almost certainly occur long before computer chips have reached the power to fulfill some of the most sought-after goals in computer science, such as the creation of extremely sophisticated electronic "brains" that will enable robots to perform on a par with humans in intellectual and cognitive tasks.

The extraordinarily small size of molecular devices brings advantages beyond the simple ability to pack more of them into a small area. To grasp these important benefits requires an understanding of how the devices work—which in turn demands some knowledge of how electrons behave when con fined to regions as small as atoms and molecules.

Free electrons can take on energy levels from a continuous range of possibilities. But in atoms or molecules, electrons have energy levels that are quantized: they can only be any one of a number of discrete values, like rungs on a ladder. This series of discrete energy values is a consequence of quantum theory and is true for any system in which the electrons are confined to an infinitesimal space. In molecules, electrons arrange themselves as bonds among atoms that resemble dispersed "clouds," called orbitals, The shape of the orbital is determined by the type and geometry of the constituent atoms. Each orbital is a single, discrete energy level for the electrons.

Even the smallest conventional microtransistors in an integrated circuit are still far too large to quantize the electrons within them. In these devices the movement of electrons is governed by physical characteristics—known as band structures—of their constituent silicon atoms. What that means is that the electrons are moving in the material within a band of allowable energy levels that is quite large relative to the energy levels permitted in a single atom or molecule. This large range of allowable energy levels permits electrons to gain enough energy to leak from one device to the next. And when these conventional devices approach the scale of a few hundred nanometers, it becomes extremely difficult to prevent the minute electric currents that represent information from leaking from one device to an adjacent one. In effect, the transistors leak the electrons that represent information, making it difficult for them to stay in the "off" state.

Besides enabling molecular devices to contain their electrons more securely, quantum mechanical phenomena can also be exploited in specially designed molecules to perform other functions. For example, to construct a "*wire*" we need an elongated molecule through which electrons can flow easily from one end to the other. Electrons in any quantized structure such as a molecule tend to move from higher- to lower-energy levels, so in order to channel electrons we need a molecule that has an empty low-energy orbital that is dispersed throughout the molecule from one end to the other.

An active device such as a transistor, however, has to do more than merely allow electrons to flow—it has to somehow control that flow. Thus, the task of the molecular device engineer is to exploit the quantum world's discrete energy levels—specifically by designing molecules whose orbital characteristics achieve the desired kind of electronic control. For example, with the right overlap of orbitals in the molecule, electrons flow. But when the overlap is disturbed—because the molecule has been twisted or its geometry has been otherwise affected—the flow is blocked. In other words, the key to control on the molecular scale is manipulating the number of electrons that are allowed to flow at low orbital energy by perturbing the orbital overlap through the molecule.

Already the standard methods of chemical synthesis allow researchers to design and produce molecules with specific atoms, geometries and orbital arrangements. Moreover, enormous quantities of these molecules are created at the same time, all of them absolutely identical and flawless. Such uniformity is extremely difficult and expensive to achieve in other batch-fabrication processes, such as the lithography-based process used to produce the millions of transistors on an integrated circuit.

The methods used to produce molecular devices are the same as those of the pharmaceutical industry. Chemists start with a compound and then gradually transform it by adding prescribed reagents whose molecules are known to bond to others at specific sites. The procedure may take many steps, but gradually the pieces come together to form a new potential molecular device with a desired orbital structure.

After the molecules are made, we use analytical technologies such as infrared spectroscopy, nuclear magnetic resonance and mass spectrometry to determine or confirm the structure of the molecules. The various technologies contribute different pieces of information about the molecule, including its molecular weight and the connection point or angle of a certain fragment. By combining the information, we determine the structure after each step as the new molecule is synthesized.

Benzene Rings

One of our simplest active devices was a molecule based on a string of three *benzene rings*, in which the orbitals over lapped (were conjugated) throughout. We made the connections between the benzene rings structurally weak, so that slight twists or kinks weakened or strengthened the conjugation of the orbitals. All we needed was a way to control this twisting and we would have a molecular device in which we could control current flow—a switch, in other words.

To the center benzene ring in the molecule, we added NO_2 and NH_2 groups, projecting outward from the string on opposite sides of the center ring. This asymmetrical configuration left the molecule with a strongly perturbed electron cloud. That asymmetric, perturbed cloud in turn made the molecule very susceptible to distortion by an electric field: applying an electric field to the molecule twisted it. We now had an active device: every time we applied a voltage to the molecule, an electric field was set up that twisted the molecule and blocked current flow. With the voltage removed, the molecule sprang back to its original shape, and the current flowed again. In follow-up experiments, we found that for our infinitesimal device the abruptness of the switching from one state to the other was superior to that of any comparable solid-state device.

Of course, a lot of advanced technology and years of research were necessary before we could even test one of these devices. The basic challenge is reaching into an unfathomably Lilliputian domain in order to contact and interact with a single molecule and bring information about the behaviour of that molecule into our macroscopic world.

Scanning Tunneling Microscope

The task was all but impossible before the invention, in the 1980s, of the *scanning tunneling microscope* (STM) at IBM's research laboratories in Zurich. The STM gives scientists a window on the atomic world, letting them visualize and manipulate single atoms or molecules. With an atomically sharp tip of metal held precisely over a surface, the topography of the surface is sensed by the minute current of tunneling electrons that flows between the surface and the tip. Rastering the tip back and forth creates a picture of the hills and valleys on the surface.

Although scanning tunneling microscopy is crucial for testing and constructing individual devices, any useful molecular circuit will consist of vast numbers of devices, orderly arranged and securely affixed to a solid structure to keep them from interacting randomly with one another. Progress toward solving this huge challenge has emerged from studies of self-assembly, a phenomenon in which atoms, molecules or groups of molecules arrange themselves spontaneously into regular patterns and even relatively complex systems without intervention from outside.

Once the assembly process has been set in motion, it proceeds on its own to some desired end. In our research we use self- assembly to attach extremely large numbers of molecules to a surface, typically a metal one. When attached, the molecules, which are often elongated in shape, protrude up from the surface, like a vast forest with identical trees spaced out in a perfect array.

Researchers have studied a variety of self-assembly systems. Our work often requires us to attach molecular devices to a metal (usually gold) surface. So we frequently work with a molecular fragment that we attach to one or both ends of our device and that has a high affinity for gold atoms. The specific fragment we commonly use, called a "sticky" end group for obvious reasons, is based on an atom of sulfur and is known in chemical terminology as thiol.

To initiate the self-assembly, we need only dip a gold surface into a beaker. In solution in this container are our molecular devices, each with thiol end groups on both ends. Spontaneously and in unimaginably large numbers, the devices attach themselves to the gold surface.

Handy though it is, self-assembly alone will not suffice to produce useful *molecular-computing systems*, at least not initially. For some time, we will have to combine self-assembly with fabrication methods, such as photolithography, borrowed from conventional semiconductor manufacturing. In photolithography, light or some other form of electromagnetic radiation is projected through a stencil-like mask to create patterns of metal and semiconductor on the surface of a semi conducting wafer. In our research we use photolithography to generate layers of metal interconnections and also holes in deposited insulating material. In the holes, we create thc clcctrical contacts and selected spots where molecules are constrained to self-assemble. Thus, the Final system consists of regions of self- assembled molecules attached by a mazelike network of metal interconnections.

Many challenges must be overcome to make a molecular device that operates analogously to a transistor. A transistor has three terminals, one of which controls the current flow between the other two. Effective though it was, our twisting switch had only two terminals, with the current flow controlled by an electrical field, In a field-effect transistor, the type in an integrated circuit, the current is also controlled by an electrical field. But the field is set up when a voltage is applied to the third terminal.

Computer Chips

A three-terminal molecular device will make possible the chemical synthesis of tremendously efficient and complex circuits. Even before then, combinations of molecular systems with conventional electronics will probably be used in places where the advantages of self-assembly are natural. But interfacing between the molecular and microelectronic worlds will present its own challenges. *Computer chips* today have two levels of size scale. From the macroscopic level of the chip we can see and hold in our hand, there is a factor of 1,000 in size reduction to get to the gross wiring level, encompassing the largest connections on the chip, which are smaller than a human hair. Then another factor-of-1,000 reduction is necessary to get to the level of the smallest connections and components of the transistors. If molecular devices are to be added to a chip, they will represent yet another factor of 1000 reduction in scale clown from the smallest microelectronic device components.

Thermal challenges are also staggering, especially if engineers wind up with no alternatives to using molecular devices in modes and configurations similar to those used now with transistors in conventional chips. At present, a state-of-the-art microprocessor with 10 million transistors and a clock cycle of half a gigahertz (half a billion cycles per second) emits almost 100 watts—greater, in radiant heat than a range-top cooking surface in the home. Such a unit is close to the thermal limitation of semiconductor technology Knowing the minimum amount of heat that a single molecular device emits would help put a limit on the number of devices we could put on a chip or substrate of some kind.

This fundamental limit of a molecule, operating at room temperature and at today's speeds, is about 50 picowatts (50 millionths of a millionth of a watt). That figure suggests an upper limit to the number of molecular devices we can closely aggregate: it is roughly 100,000 times more than what we can now do with silicon micro transistors on a chip. Although that may seem like a vast improvement, it is still far below the density that would be possible if we did not have to worry about heat.

Right now no one knows how to create such an interconnect structure on the molecular level. Straightforward extensions of the present techniques we employ to fabricate complex micro electronics are not practical for molecular-scale electronics, because the lithography needed for creating the interconnections to single molecules is far beyond the capability of known technologies. Is the ability to address every device, the common architecture we use today necessary or efficient at molecular-scale densities? What will large-scale circuits of this technology look like? Can we use nano tubes, single-walled structures of carbon with diameters of one or two nanometers and lengths of less than a micron, as the next generation of interconnects between molecular-scale devices?

Decades from now, radical departures from present computing design will probably be needed to exploit molecular computing systems fully if we are to extend electronics significantly beyond Moore's Law. We have only very limited ideas about what these departures might be. The ability to construct complex molecular devices, with new paradigms and lists of rules about connecting the various devices, will open up an entirely different way to think about computer design.

Although such departures are fraught with problems, we have no alternative but to solve them if electronic is to continue advancing at something like its current pace well into the next century. And difficult though the challenges may be, the rewards for those who solve the problems could be staggering. By pushing Moore's Law past the limits of the tremendously powerful technology we already have, these researchers will take electronics into vast, uncharted terrain. If we can get to that region, we will almost certainly find some wondrous things—may be even the circuitry that will give rise to our intellectual successor.

13

Technological and Scientific Aspects

Since two decades *nanotechnology* has evolved from different scientific fields, such as physics, chemistry, molecular biology, and material science. The nanotechnology aims to study and to manipulate real-world structures with sizes ranging between a nanometer, i.e. one millionth part of a millimeter, and up to one hundred nanometers. The set of typical "*nano*"-objects includes colloidal crystals, molecules, DNA based structures, and integrated semiconductor circuits.

The first scientist who pointed out that "there is plenty of room at the bottom" was Richard Feynman in the year 1959. He envisioned scientific discoveries and new applications of miniature objects as soon as material systems can be assembled at the atomic scale. To this end, machines and imaging techniques would be necessary which can be controlled at the nanometer or subnanometer scale. Fifty years later, the scanning tunneling microscope and the atomic force microscope are ubiquitous in scientific laboratories allowing to image structures with atomic resolution. As a result, various disciplines of the nanotechnology aim towards manufacturing materials for diverse products with new functionalities.

There are two strategies for assembling nanosystems: the *top down* and the *bottom up* approach. The "top down" approach follows the development of the microelectronic industry to miniaturize integrated semiconductor circuits. Modern lithographic techniques enable to pattern nanoscale structures, such as transistor circuits, with highest precision down to several nanometers. In 1965 Gordon Moore, co-founder of

Intel corporation, predicted that the number of transistors on a computer chip would double about every eighteen months. The exponential law, also known as *Moore's first law*, has described the development of integrated circuits surprisingly well for decades. As the market for information technology continues to grow, the demand for computer hardware instigates more and more sophisticated "top down" techniques to build more densely packed transistor circuits. *Moore's second law* states that the implementation of a next generation of integrated circuits at minimum cost will be exponentially more expensive as well. Until all the constraints will finally limit the growth of the semiconductor "*top down*" industry, scientists and engineers assume that the nanotechnology will give answers to most of the technological challenges. For instance, as soon as the feature size of the semiconductor transistors reaches the level that quantum phenomena are important, different concepts for the assembly need to be considered. One possibility is the "bottom up" approach, which is based upon molecular recognition and chemical self-assembly of molecules. In combination with chemical synthesis techniques the "bottom up" approach allows assembling of macromolecular complexes with new functionalities.

Nanoscience and *nanotechnology* will definitely have a strong impact on our lives in many separate areas; e.g. information technology, material sciences, and medicines, to name only a few. The manuscript starts with a paragraph about the most recent developments of lithographic techniques in industry and nanoscience. Moreover, most of the discoveries of the nanotechnologies were initiated by the development of accurate microscopes with atomic resolution. Therefore, the manuscript also gives a short introduction to scanning microscopes. A paragraph about nanoelectronics features approaches and techniques which are supposed to produce successor technologies of the microelectronic industry. Examples include quantum computing and spintronics, two emerging fields of nanoscale electronic circuits. Alternative materials and approaches are currently being investigated for novel products consisting of nanostructures. Aspects of nanoscale materials as well as the impact of the nanotechnology on health sciences are finally described in the manuscript.

Top-down Technique

The optical "*top down*" lithography has been the backbone of the semiconductor industry since 45 years. In this technique a so-called *photoresist*—a liquid, photosensitive chemical that resists etching processes—is spin-coated onto the polished surface of a semiconductor wafer. After hardening of the photoresist, e.g. by heating, the semiconductor sample with the photoresist layer on top is exposed to light. Hereby, patterns can be defined in the photoresist. After a developing process, the exposed (or the unexposed) parts of the semiconductor wafer are bared. By follow-up processes such as etching, metallization, and oxidation the patterns can be translated to conductor paths, logical circuits, or memory cells. Since the eighties of the last century, the demise of the optical lithography has been predicted as being only a few years away. However, each time the optical lithography has reached a limitation, new techniques extended the economically useful life-time of the "*top down*" technique.

The minimum feature size of optically defined patterns depends on the wavelength of the utilized light as well as factors, which are due to e.g. the shape of the lenses and the quality of the photoresist. In 2003, the line-width of semiconductor circuits fell below one hundred nanometers; i.e. the semiconductor industry can be literally seen as being part of the nanotechnologies. For this achievement, Argon Fluoride Excimer lasers are applied with an optical wavelength of 193 nm in the deep ultraviolet. For lithography in this optical range tricks such as *optical proximity correction* and *phase shifting* were invented and successfully implemented. On one hand, the miniaturizing of semiconductor circuits is limited by economic costs for the semiconductor industry, since the implementation of new techniques to realize an ever smaller feature size results in ever increasing costs. On the other hand, physical material properties, such as the high absorption level of refractive mirrors at short optical wavelengths, give a natural limit of only a few tens of nanometers for the miniaturization process. Assuming Moore's laws, the final limit for the optical lithography is supposed to be reached in less than a decade.

At present, there are several possible "top down" successor nanotechnologies: e.g. the *extreme ultraviolet light technique* (EUV), the electron beam lithography with multicolumn processing facilities, the *focused ion beam technique* (FIP), and last but not least the ultraviolet nano-imprinting technique. The implementation of each of the above techniques implies vast technical challenges to overcome. The most promising technique is the one utilizing extreme ultraviolet light of a wavelength of only thirteen nanometers. For this technique, fabrication errors of the "optical" components need to be in the nanometer or subnanometer range. For comparison, the state-of-the-art x-ray telescopes, such as the *XXM Newton* telescope of the Zeiss AG, exhibit a granularity of the mirror surfaces of 0.4 nm. In addition to the "state of the art" optical requirements, all metrological components of the extreme ultraviolet technique need to exhibit subnanometer resolution. As a result, the costs of a stepper machine, by which photoresists in the extreme ultraviolet can be exposed, are fifty millions dollars per system; enabling a line-width of 35 nanometers.

For medium-sized businesses, the *ultraviolet nano-imprinting technique* seems to be the most promising method to fabricate nanoscale circuits. Here, nanostructures are mechanically imprinted into a photoresist. The stamp with the nanoscale patterns is made out of fused quartz; a material which is transparent for ultraviolet light. As soon as the stamp is plunged into the photoresist, a short ultraviolet light pulse causes the photoresist to polymerize according to the patterns on the stamp. In line with the optical lithography, follow-up processes allow defining nanoscale circuits in various geometries. The minimum feature size of the nano-imprinting technique is about ten nanometers. At the same time, the technique is applicable to metals and plastic materials. The costs for an industrial nano-imprinting machine are supposed to be less than one million Dollars. However, the throughput of nano-imprinting machines is much lower than for stepper machines.

Bottom-up Technique

The antipole of the "top down" approach is the so-called *bottom up technique*. Generally, "bottom up" assembly techniques seek to fabricate composite materials comprising of nanoscale objects which are spatially ordered via molecular recognition. The primary examples of the technique are *self-assembled monolayers* (SAMs) of molecules. A substrate, usually made out of metals, is immersed into a dilute solution of a surface-active organic material that adsorbs onto the surface and organizes via a self-assembly process. The result is a highly ordered and well-packed molecular monolayer. The method can be extended towards *layer-by-layer* (LBL) assembly; by which polymer *light-emitting devices* (LED) have already been fabricated. The self-assembly technique also allows positioning of single molecules in between two metal electrodes, and subsequently into an experimental circuit. By this setup, quantum mechanical transport characteristics of single molecules, such as photochromic switching behaviour, can be studied in order to build electronic devices with new functionalities.

Monolayer Techniques

There are several combinations and variations of the self-assembled monolayer techniques. It can be combined with nano-imprinting methods, the atomic force microscope, or the focused ion beam technique; allowing the fabrication of geometrical patterns of molecules with varying friction, chemical functionality, and/or topological characteristics. Since most of the processes are performed in solution, electrical fields give further possibilities to assemble charged compounds in a directed way; e.g. for creation of functionalized sensing electrodes. Electrodes modified by negatively charged gold nanoparticles, and positively charged host molecules have proven highly sensitive for the detection of e.g. adrenaline (as the guest molecule). A further very promising field is the DNA directed assembly of network materials. Here, molecular recognition reactions between single DNA strands are translated into aggregate formation of other materials such as nanoparticles. For instance, the molecular recognition of DNA molecules has been exploited to build so-called nano-tweezers. Last but not least, material networks can be built by so-called block copolymer templates. A typical application of this technique is the formation of networks of metallic nanowires; i.e. metals are vapour-deposited onto a preformed template matrix made out of copolymers. The polymer networks can be used as two- or even three-dimensional templates, while the glass transition of the polymers gives further flexibility to form such porous films.

Generally, there is a wide range of materials which can be lithographically engineered by "bottom up" techniques. The corresponding research has led to numerous sensing, electronic, optoelectronic, and photoelectronic interfaces as well as devices. Prominent examples of nanostructures, which can be assembled with "bottom up" techniques, are presented below in the paragraph entitled "*material sciences*".

Scanning Microscopes

Scanning microscopes are key inventories for the nanotechnology. On one hand they allow probing the characteristics of nanoscale objects with highest resolution. Examples include topology, material configuration, electrical, chemical, magnetic, and optical properties of the studied objects. On the other hand, scanning microscopes allow local manipulation of the nanostructures. In a seminal work in 1982, Binnig and Rohrer invented the *scanning tunneling microscope* (STM). Piezo crystals move a scanning tip across the surface of a sample, while the electric current is recorded between the tip and the sample. If the tip is located very close to the surface of the sample, the electric current comprises of tunneling electrons. To simplify matters,

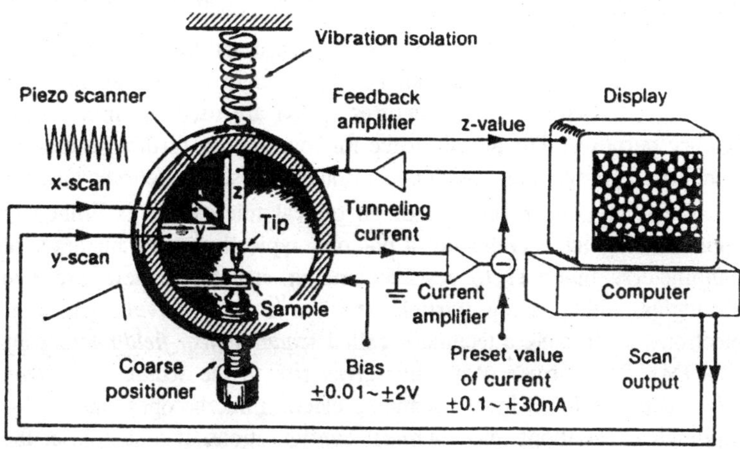

Fig. 13.1. Scanning tunneling microscope.

electrons behave as waves in the quantum world. As soon as the distance between the tip and the surface is in the order of the electron wavelength, electrons can tunnel from the tip to the surface. The quantum nature of the tunneling process ensures that the point of the tip, which is closest to the surface, contributes mainly to the current. In principle, this point can be made up of only one atom, which allows atomic resolution. The scanning tunneling microscope is sensitive to electronic densities on the surface of the sample, which allows imaging of electronic orbits of atoms. At the same time, the scanning tunneling microscope can operate at an atmospheric pressure down to high vacuum conditions at various temperatures, which makes it a unique imaging and patterning tool for the nanosciences. E.g. scanning tunneling microscopes are utilized to pattern nanostructures by moving single atoms across surfaces, while the corresponding change of the quantum mechanical configuration can be recorded *in situ*.

The atomic force microscope operates similarly to the scanning tunneling microscope. Here, the force between the scanning tip and the sample surface is extracted by measuring the deflection of the tip towards the sample. Again the atomic force microscope can be utilized as an instrument to image and to manipulate structures on the nanometer scale. Most importantly, the atomic force microscope is a unique tool to measure forces between two nanoobjects with a resolution of only a few PikoNewtons (a millionth of a millionth Newton). By functionalizing the scanning tip chemically, a single molecule can be utilized for sensing applications; e.g. the binding forces between ligands and receptors can be determined for biological and health applications. Furthermore, magnetic tips can be exploited to image the magnetic topology of magnetic nanostructures which are used for information science purposes. The atomic force microscope can further be used as a miniature groove, e.g. which cuts through nanoscale electronic circuits.

Generally, the resolution of an optical microscope is limited to approximately the wavelength of photons; typically several hundreds of nanometers. In order to resolve smaller structures there are two strategies, either using particles with a shorter wavelength, e.g. electrons, or utilizing a technique called *scanning near-field microscopy* (SNOM). The former possibility gives rise to the *scanning electron microscope* (SEM). Modern scanning electron microscopes can resolve topological variations and chemical configurations of a sample at the subnanometer length scale. To this end, a sample is located in a high-vacuum chamber, and a beam of electrons is focused onto the surface of a sample. The reflected electrons are then detected as a function of the position of the initial beam. In order to study the optical properties of nanoscale structures, which are substantially smaller than the photon wavelength, the *scanning near-field microscopy* (SNOM) is an excellent method. Here, glass fibers with apertures of only hundred nanometers or less are utilized. The size of the aperture defines the spatial resolution of the technique, while the intensity of the gathered light is recorded as a function of the position of the glass fiber.

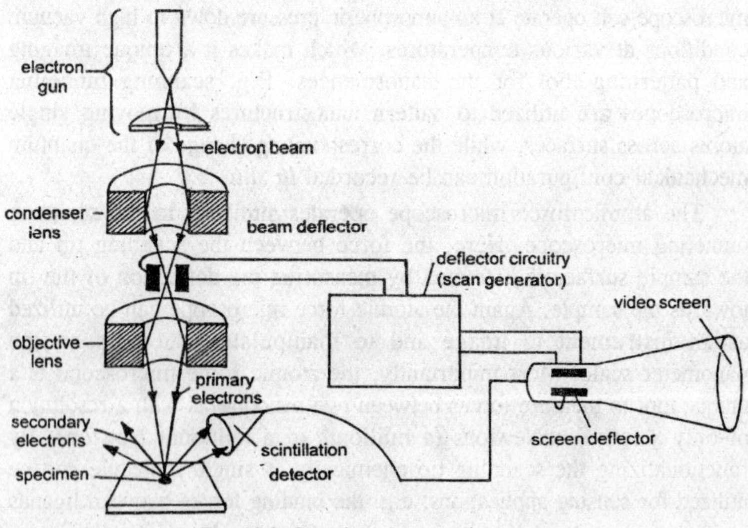

Fig. 13.2. Scanning electron microscope.

Nanoelectronics

The field of *nanoelectronics* aims towards developing devices comprising of nanostructures with new overall functionalities for the information technology. By exploiting electronic, optical, and magnetic properties of nanostructures, integrated circuits and memory devices are developed for faster and more reliable information processing schemes at lower heat dissipation and with a greater portability of future computers. Since the seventies of the last century, so-called quantum wells in semiconductor heterostructures have been subject of intensive scientific research. Heterostructures are semiconductor crystals with composite monolayers of different atoms. Most importantly, heterostructures can be grown in which electrons are confined solely to the plane of the *quantum well*. These investigations on nanoscale monolayers of composite semiconductors yielded to applications such as the CD drives, the laser printers, and the input-amplifiers of cell phones; investigations which were honored with the Nobel Prize in 2000.

Since approximately fifteen years, so-called nanocrystals or colloidal quantum dots are promising compounds of a successor nanoelectronic technology. Generally, the colloidal quantum dots are semiconductor crystals in which an electronic quantum state is localized within a tiny volume. Most importantly, the optical and the electronic properties of quantum dots are defined by the quantum mechanical characteristics of the electron state. A nanoelectronic device, which is very similar to a quantum dot, is the so-called *single electron transistor* (SET). The transistor relies on the switching of a single electron. The advantage of such devices would be the combination of minimum size with a minimum power dissipation, both of which become very important issues for densely packed logical circuits.

At the same time, optical processing schemes are expected to become important parts of the future information technology. Only recently Raman lasers, optical waveguides, and fast optical switches have been realized on silicon chips. The realization of an all-silicon based optical processing scheme is an important step to combine fiber optics with the present information technology at low cost. So far optical signals in fibers can be switched by micro-mechanical elements, and/or optical information is translated into electronic signals e.g. by photo diodes. Subsequently, the information is processed via electronic transistor elements and in turn, it is transferred again to optical information by a laser. This very inefficient scheme is expected to be replaced by the silicon-based technology mentioned above. In future devices, further nanostructures, such as *photonic crystals*, are expected to play an important role to control light within a semiconductor crystal directly. In "photonic crystals" the optical constants of a material are modulated in three dimensions by nanostructuring the (semiconductor) host crystal. As a result, only photons with a defined wave vector and energy are allowed within the periodic structure. The incorporation of nanoscale defects into a photonic crystal further allows the deflection of certain optical wave modes within the photonic crystal. Moreover, photonic crystals can be designed, in which well-defined optical modes can interact with nanostructures. Recently it was demonstrated that by positioning a quantum dot at the center of an artificially designed optical mode volume, the electron states of the quantum dot strongly interact with the electromagnetic fields of photons. Latter experiments give rise to the feasibility of an information processing schemes purely based on quantum mechanics.

In principle, the atomic *granularity* defines a natural limit to the miniaturization of electronic circuits. The electron orbits of atoms are the smallest building blocks of a possible "nano-computer". In other words, a conductor path can not be narrower than a single atom. Therefore, it is reasonable to ask whether a computer, which exploits the quantum nature of the electron states, is technically feasible. The quantum mechanical tunneling effect is one of the more obvious effects which will arise as soon as the dimensions of the electronic circuits fall below a certain value. Since electrons are waves, they may tunnel from one conductor path to another if the paths are located in a distance which compares to the wavelength of the electrons.

Typical electron wavelengths in semiconductor crystals are in the range of several nanometers. Albeit the resulting leakage currents are small, the total effect can be significant for millions of densely packed transistors on a chip. On one hand, tunneling electrons give rise to additional heat dissipation. On the other hand, dislocated charges have to be considered as logical errors and/or they influence the general functionality of the integrated circuit via the Coulomb interaction; e.g. via *cross-capacities* between adjacent conductor paths. Furthermore, an effect called *electro migration* can derogate conductor paths made out of e.g. Aluminum; i.e. the electron current *flushes away* single metal atoms.

The quantum mechanical nature of electrons comprises properties and corresponding technical possibilities which go beyond the functionality of a classical computer. In principle, the classical desk top computer is based upon semiconductor physics with foundations in quantum mechanics as well. However, the functionality of the classical computer can be simply described by "classical" physics. The quantum computer aims towards exploiting the physical laws of quantum mechanics to make certain calculations more efficient. As far as the quantum computer is concerned, the equivalent to the binary digit (BIT) of the classical computer is the so-called *quantum bit* or *qu-bit*. In quantum mechanics, particles such as electrons are described by a set of independent parameters, which are called *quantum numbers*. For instance, the magnetic moment of an electron, the so-called *spin*, is a possible quantum number, which can be mathematically represented by a plus or a minus one half. A quantum bit consists out of the sum of independent quantum numbers, while continuous prefactors define the weight of a specific quantum number. The superposition of quantum numbers also includes so-called entangled states. Latter are physical states of a quantum mechanical particle, where one quantum number depends logically on another in an intrinsic way. The entanglement is unknown to classical physics (and the most intuitive way to accept its reality is via mathematics). Most importantly, the superposition and the entanglement of quantum bits give rise to fast and parallel information processing schemes at a minimum heat dissipation.

A further discipline of nanoelectronics is the field of spintronics, where both charge and spin degrees of freedom of an electron are exploited to realize new electronic devices. As mentioned above, the "spin" describes the magnetic moment of an electron. It is classified by two values, which are usually referred to as "up" and "down" (the two expressions depict the fact that the magnetic moment of an electron orientates in the directions up or down with respect to an external magnetic field). The magnetic interaction of an electron spin with the crystal environment is much lower than the Coulomb interaction felt by electron charges. Therefore, heat dissipation effects are predicted to be significantly smaller for "spintronic" circuits than for "electronic" circuits. However, there are ubiquitous spin relaxation mechanisms in semiconductor crystals, by which the orientation of the spin (and thus the corresponding classical information) is lost within several hundreds of nanoseconds.

The precise fabrication of nanoscale crystalline layers has also given rise to the development of magnetic layer systems, the resistance of which alters significantly when small magnetic fields are applied.

Latter effect is known as the *giant magnetoresistance* (GMR). The technology has produced the prime example of a spintronic device: the so-called *magnetic random access memory* (M-RAM). The memory concept was originally proposed by IBM, and it is now introduced into the marked of information storage technology. In this technology, the information is stored by magnetic domains, while the orientation of the domains is read out via the *giant magnetoresistance*. In principle, the M-RAM technology combines all advantages of the present memory technologies: it is a fast, nonvolatile information storage technology, and it allows information storage at high density and minimum heat dissipation. The magneto-mechanical hard drive, present in most of the current computers, provides also a medium for high density, nonvolatile information storage. However, it is comparably slow. Hard drives, based upon a concept called *dynamic random access memory* (D-RAM), are fast, however, they need to "refresh" the information iteratively by current pulses. The "flash" memories, which are utilized in MP3-players, cell phones, and digital cameras, are quite slow and they can only be used approximately one million times. Therefore, M-RAM devices seem to be promising candidates for a "*nanoelectronic*" successor technology.

Last but not least, nanomaterial sciences give further possibilities for new architectures of integrated circuits and memory devices. For instance, a memory concept, usually referred to as *phase change random access memory* (phase change RAM), utilizes the phase transition between "*crystalline*" and "*amorphous*" states of materials to encode information. The phase transition is triggered by electrical impulses, while the resistance difference between the crystalline and the amorphous state allows identifying the binary information. The crystalline state is re-initialized, if the material is homogeneously melted by the application of long electrical current. In principle, the method allows nonvolatile information storage at highest densities; i.e. a Terabit on the area of a stamp. A corresponding method is already implemented in *digital versatile disks* (DVDs) made out of polymers, where a short laser pulse melts small areas of the disk; i.e. the spots undergo a phase transition from crystalline to amorphous. Generally, for memory device with highest information densities, material sciences will play an important part in nanoelectronic engineering.

Material Sciences

In recent years, the material science and technology have explored a variety of new materials in which nanoscale components give rise to better properties of well known products. To this end, optical, electric, thermal, mechanical, and chemical characteristics of nanostructures are exploited to enhance certain properties of the composite systems. Examples of "nano"-products include paints, inks, cosmetics, lubricants, grinding pastes, ceramics, luminescent materials, glues, and protective lacquers; to name only a few. The new materials are envisioned to fulfill tasks, such as the ability to decompose pollutants at higher rates or to converse light into current more efficiently. For such and more complex tasks novel materials are based on several nanoscale components whose spatial organization is engineered at the molecular level. The macroscopic behaviour arises from the combination of the novel properties of the individual building blocks and their mutual interaction.

Materials with nanoscale pores, such as zeolites, are utilized for catalyzers in chemical processes or as ten-sides in cleaning detergents. Metallic alloys show a certain degree of mechanical memory. New terms such as nanotubes, nanowires and quantum dots are now common jargon of scientific publications. These objects are among the smallest, man-made units that display physical and chemical properties which make them promising candidates as fundamental building blocks of novel transistors.

Quantum dots are the ultimate example of a solid, in which all dimensions shrink down to a few nanometers. The electronic and optical properties of quantum dots are a consequence of their dimensions. At the same time, colloidal quantum dots can be synthesized from a wide range of materials. The dimension of these particles makes them ideal candidates for the nano-engineering of surfaces and the fabrication of functional nanostructures. Moreover, semiconductor quantum dots are probably the most studied nanoscale systems at the moment.

The surface of the lotus flower is *hydrophobic* due to nanoscale structures. In a biomimetic ansatz, these properties have been transferred to paints which are strongly "*soil resisting*". Investigations of the nanoscale architecture of shells and bones gave rise to new developments in the field of layered materials which are extremely robust. Carbon nanotubes have the topology of a graphite sheet which is rolled up to form a tube. These wires with diameters of only a few nanometers and a controlled chemical make-up show intriguing electrical properties and extremely high strength and stiffness. Due to their properties, carbon nanotubes are proposed to be utilized e.g. as field emitters in flat screens. At the same time, carbon nanotubes exhibit a unique ratio of inner volume to surface, which makes them ideal candidates for extremely light hydrogen storage devices. Moreover, compounds made out of carbon fibers are light and robust materials which are used in a broad range of application.

An important technique of the material sciences is the sol/gel technique, by which nanostructures are formed and controlled by the application of so-called *colloids*. Generally, colloids are systems in which droplets of one substance are formed and in turn, resolved within a second substance. Daily life examples of colloids are the "sauce bearnaise", in which droplets of vinegar are resolved in butter. Other examples are cosmetic cremes and paints. In material sciences colloids are exploited to form nanoscale structures with chemically controlled composition and/or nanoscale structures which tend to build networks as soon as the solvent is removed.

By utilizing the sol/gel technique, protective lacquers of a variety of substances have been produced. E.g. silicon nanostructures can be assembled in the liquid phase, and in turn, they are sprayed onto the surface of arbitrary surfaces. After the solvent has vapourized, appropriate heating gives rise to the formation of a ceramic cover layer which is extremely thin and hard at the same time. Due to the large surface to volume ratio of nanostructures, such ceramics can be formed at rather low temperatures. The sol/gel technique can be even utilized to fabricate optical components such as glass fibers and frequency doublers.

A whole class of nanomaterials is defined by so-called *aerogels*, which are highly porous materials consisting mainly out of "air". Again, a daily life example of such a material can be found in the French cuisine. The "Baizer" consists of white egg in which thousands of microscale air bubbles are incorporated. The air bubbles in "*Baizer*" have a size of several microns. As a result, visible light is scattered at the air inclusions and "Baizer" appears as a white substance, while white egg is transparent to visible light. If the enclosed air bubbles have a submicron diameter, aerogels appear to be transparent.

At the same time aerogels are very good heat insulators, since heat can not circulate in the air bubbles. Hereby, window glasses with remarkable insulator properties can be fabricated. Aerogels with nanometer inclusions can be produced by the sol/gel technique; i.e. spheres are formed within a colloid by utilizing networking nanostructures. A gel is formed as soon as the solvent is removed while the volume of the sol is kept constant.

The large inner surface of an aerogel gives rise to several applications in the energy sector. Lithium batteries with enhanced storage characteristics have been built. Electrical capacitors of up to 2500 Farads have been fabricated by aerogels. Last but not least, better fuel cells can be envisioned by this technique.

Health Sciences

There are several applications of *nanotechnologies* which are already implemented in commercially available cosmetic products. For instance, nanoscale spheres, made out of apatites and proteins, are implemented into toothpastes in order to enable *biomineralization* of the teeth, because the corresponding nanospheres consist out of the same material as teeth. Other well-known examples are sunscreens which contain colloids of zinc oxide. Such a sunscreen gives an excellent protection against ultraviolet radiation, which is simply reflected by the nanoscale structures, while the sunscreen appears transparent to human eyes. In addition, aerogels can be utilized for a very efficient drug delivery. Nanoscale capsules with functionalized molecular linkers can be utilized to deliver drugs to metastases via molecular recognition.

Recently, a method called *magnetic fluid hyperthermia* was introduced to oncology, which is based on heating tissues for therapeutic purposes. Generally, tumour cells exhibit a hypersensibility to heating (*hyperthermia*). At the same time, surfaces of magnetic nanoparticles can be functionalized in a way that they accumulate only/mainly in tumour tissue. Therefore, if an alternating magnetic field is applied, the dissipation losses, induced due the movement of the magnetic colloids, heat up and, in turn, destroy the tumour cells. By solving the biochemical and physiological specificity problem, cancer-specific hyperthermia protocols have been developed.

A well-known vision of scientists working in the field of nanotechnology is the "lab on a chip". The chips would have a size of roughly a square centimeter, while millions of nanoscale instruments would analyze droplets of liquids for biological, medical or forensic purposes. One possibility to move the liquids across the surface of the chip is given by so-called *surface acoustic waves* (SAW). The latter technology has recently been realized for commercially available products.

To conclude, nanoscience and nanotechnology have a great potential to solve several challenges of the modern society. Moreover, there exists a great economic expectation for growth related to nanoscale products. At present, there are about 450 companies alone in Germany, which manufacture products related to nanomaterials. To this end, the manuscript intends to give an introduction to some of the nanotechnologies which will become important to the daily life in the near future; in particular these are the information technology, material sciences, and health sciences.

14

MOLECULAR MANUFACTURING

Molecular manufacturing will relax many constraints regarding materials strength, cost, and reliability. Many capabilities that will become possible with these relaxed constraints are capabilities that people have wanted for many years, and discussed back in the 1950s and 1960s. Because of this, a common reaction to descriptions of such capabilities is that they sound like stale futurist projections, or like old science fiction. Because the old predictions did not come to pass, people are prone to reject similar predictions made now. What is different this time is that we are facing a profound transformation in our technology base that will make many new developments possible, including new developments that follow old and familiar ideas. One of these is opening the space frontier.

Foundational technologies are already discussed, pathways forward from those technologies, and advanced nanotechnology based on molecular manufacturing and assemblers have already been discussed. Advanced capabilities, the ability to produce large quantities of diamond-based structural materials (and related atomically precise components) at costs that approach raw material costs are taken into consideration.

OUR HISTORY IN SPACE

Once upon a time, there was the *V-2 rocket*, then longer-range *ballistic missiles*, and then *orbital rockets* derived from ballistic missiles. These are still in use today. Orbital launches today, like V-2 launches during World War II, depend on chemical rockets burning fuel to produce a fast stream of exhaust. A basic rule regarding rockets is that it is easy for a rocket to reach a speed comparable to that of its exhaust, but exponentially more costly to go beyond that speed (more and more fuel must be burned to accelerate fuel, and the cost grows explosively). Earth escape velocity is more than twice the speed of chemical rocket exhaust.

The space program today seems trapped, forever flying expensive, unreliable vehicles to Earth orbit with occasional robotic missions to the planets. And it remains too much a space *program*—fortunately, we do not have a similar air, land, or sea program, Why are we trapped? Chiefly because of our manufacturing technology base. If we could make structures with a far higher strength-to-weight ratio, a far lower cost per kilogram, and far higher precision and reliability, space would no longer seem like a science fiction realm. Instead, it would be a part of the world in which we live, and Earth would be seen as a very small part of that world.

Better Vehicles to Get to Space

Much can be done to improve *chemical rockets*, chiefly by making the vehicle stronger and lighter so that it can hold more fuel while needing less fuel to accelerate its own weight. So far, these vehicles have been made of *aerospace aluminum*, which has less than 2 percent of the strength of an equivalent mass of diamond-based structural material. Put another way, if one makes two structures of equal size and strength from these two materials, the aluminum structure must be more than 50 times as massive. In spacecraft, mass is bad, and a shift to materials based on diamond (or related materials) can be very good, indeed.

With better structural materials, spacecraft performance can be absurdly good. If you use materials with the strength- to-weight ratio of diamond to build a rocket fueled by liquid oxygen and liquid hydrogen, you can build a small, single-stage vehicle that takes four people to orbit. With strong materials, wings work well, and the vehicle can take off and land on an ordinary runway (you can afford to carry large wings to orbit if they do not weigh much). Since the vehicle is made of strong, hard materials that do not undergo appreciable wear or fatigue during a flight, it is ready to fly again once you have refueled it.

The mass of such a vehicle need not be larger than 3 metric tons—the mass of a heavy station wagon. That mass includes fuel. If you remove the fuel, the passengers, and the passenger cabin, the mass of the vehicle need not be more than 60 kilograms. If you were to make a similar structure of equivalent strength using aluminum, the mass of the vehicle would be greater than the mass of its fuel. You cannot build spacecraft that way, and the spacecraft that we use today have much more efficient structures than the simple design used in this exercise. Sixty kilograms is considerably more structural mass than is necessary in a good design, providing a large margin of safety.

Getting Around in Space

Several propulsion systems expected to work well in space have not been pursued by NASA in its last 20 years of bold innovation. One class of systems includes sails driven by the *pressure of sunlight*. The magnitude of this pressure equals the power density divided by the speed of light, times 2 for a perfect reflector face on the light; at Earth's distance from the Sun, this is roughly $2 \times (1{,}400\ \mathrm{W/m^2})/(3 \times 10^8\ \mathrm{m/s}) \approx 9.3 \times 10^{-6}\ \mathrm{N/m^2}$, or 9.3 N per square kilometer. A sail consisting of many panels of 100-nanometer-thick aluminum foil (about 0.27 $\mathrm{gm/m^2}$), all supported on a low-mass tension structure, will experience enough pressure from incident sunlight (within the inner solar system) to accelerate a spacecraft by kilometers per second per day, depending on the mass and the payload being carried. The sail can do this without throwing away any reaction mass. By tilting the reflector to use light pressure to decrease orbital speed, the sail can use solar gravity to move closer to the Sun. Sails can carry space craft around the inner solar system quite quickly, compared to chemical rockets.

Competing with solar sail ideas back in the 1970s was the concept of solar electric propulsion. This works by using sunlight to generate electricity, which then (typically) powers an ion engine to expel charged particles at high speeds, much faster than the exhaust of a chemical rocket. All else being equal, the velocity that can be attained by a rocket is proportional to the velocity of its exhaust. Chemical rockets have exhaust velocities of around 5 km per second. A typical exhaust velocity for ion engines optimized for today's technology is 100 km per second, but the acceleration of the vehicle is less than 1/1000 of the acceleration of Earth's gravity. For an ideal engine, these numbers vary inversely—doubling the exhaust velocity cuts the acceleration in half.

The *Journal of the British Interplanetary Society* recently did a special issue on applications of nanotechnology to space and space systems. In this issue, some calculations on solar electric propulsion based on molecular manufacturing, assuming device physics like that familiar today, but with tiny cells, tiny optical concentrators, and tiny heat-radiating fins are outlined. The key idea is to focus sunlight to

1,000 times normal intensity. This has been done in the laboratory, and actually increases the efficiency of the cell somewhat, so long as the cell is kept cool. Thus, the power output of the cell is multiplied by more than 1,000. Cooling works well in small structures because in short radiating fins, a small temperature difference creates a huge thermal gradient. The macroscopic structure of a vehicle ends up looking a lot like a light sail, and the mass-per-unit area is similar. A design exercise based on these principles indicates that the power output per kilogram can be increased from a fraction of a kilowatt (as in present spacecraft solar arrays) to more than a megawatt. This can enable solar electric propulsion systems to have enormous exhaust velocities compared to chemical rockets—around 1,000 km per second, rather than 5—while accelerating at 1 gravity.

To fly from here to the East Coast today with a stop in Chicago takes about six hours. With a solar electric propulsion system of the sort I just described, going from just outside the atmosphere over Palo Alto to just outside the atmosphere over, say, Washington, DC, could be done in about the same length of time, but with a brief stop on the Moon.

Why Go into Space?

Why is space of basic importance? Because it is almost everything. Most of the energy supply in the solar system is not in the Earth's rocks, buried as oil, coal, or uranium. It is not in sunlight striking the Earth. It is sunlight that is fleeing into interstellar space, never hitting a planet or anything else on its way out. The Sun puts out more energy in a second than the human race would use at its present rate of consumption in roughly a million years.

Space is also rich in raw materials. The material in the asteroid belt—rubble left over from the formation of the planets—is enough to bury Earth's continents kilometers deep in raw material. Humanity has used a very small fraction of the top kilometer of Earth's crust, so the amount of material that is practically avail able from these little scraps of the *Solar system* (the *asteroids*) is vastly more than the amount of material that anyone would ever want to take out of Earth's crust. That material is out there, dead, sterile, and waiting to have something done with it.

Living in Space

What would it be like to live in space? Today's space program use big, bulky, stiff, inflated space suits that are awkward and uncomfortable. With molecular manufacturing, however, one could make suits from remarkable new materials. A three-dimensional woven fabric consisting of fibers that can be actively shortened and extended by using nano mechanical motors. Smart controllers can be used to make suits that are flexible, supple, and hold atmospheric pressure. They would bend actively to conform to the body, instead of being pushed around by the wearer. With other appropriate nano mechanisms, they could supply air and keep the wearer cool and comfortable in a wide range of conditions. So, it appears possible through molecular manufacturing to have suits that actually would make it comfortable to lounge around in vacuum and to walk about on the Moon. They could even make Boston habitable in the winter!

We have little reason, though, to go jumping around in vacuum. When people think about space, most still picture little cabins in spacecraft, people in space suits, and perhaps planets and domes on planets. This imagery has been very common. In the 1970s, Gerard O'Neill launched the idea of space settlements, but this proposal ran into problems rooted in manufacturing. Space transportation was too expensive, but the physics was sound. It is possible to make cylindrical, habitable structures in space that are kilometers across. With high strength-to-weight ratio (diamondoid) materials, those kilometers could be multiplied by a factor of 10 or more. For the larger radii, rotation at less than one revolution per minute would generate one gravity of force, for a person standing in the interior.

External mirrors would capture sunlight and reflect it through transparent windows to shine upon the interior of the settlement. The interiors of such settlements could be like ***suburbia***, if that is what people want. Or, they could be much like other familiar or unfamiliar landscapes. These settlements would be made possible by access to the raw materials of space—to energy and mass—and by the use of molecular manufacturing for conversion of these materials into efficient, reliable, low-cost hardware. Using the raw materials of the asteroids alone, it will be possible to build roughly a thousand times Earth's land area, and to make living space as pleasant as any place on Earth. To be sure, there would be differences. As you look toward the horizon, the land curves up. And if you were to dig a deep hole, well, it would not be quite the same. Space settlements cannot replace Earth.

From an environmental perspective, however, if you have many people living on the Earth, and some of them care about the Earth for its own sake, but some do not, and would just like some place to hang out and do things like set off explosives and build parking lots, then it would be a good thing for the Earth for those people to have somewhere else to go. Indeed, some of those other places could provide living space that, from the point of view of animals and plants, would be perfectly acceptable. Animals and plants would not be bothered by the artificial nature of foundations of their environment, because their point of view, what matters is not abstractions such as, "Is the origin of this place *technological*, or *geological*?" What matters instead are questions such as, "What's the sunshine like?" "How is the soil?" "Is the air and water clean?" "Is my habitat going to be paved over by a freeway?" This transition in technology, if managed well, could provide new places for life, including people, in space. And by easing pressures from human beings pursuing their wants, it could make the Earth a better place for nature.

15

IMPLICATIONS OF NANOTECHNOLOGY

Nanoscale science and *engineering*, generally referred to together as *nanotechnology*, covers the investigation and manipulation of matter at the nanometer length scale, from which revolutionary technical applications are being expected. Under the umbrella of *nanotechnology* an intriguing diversity of formerly distinctive fields of science and engineering research flourishes, including physicists, chemists, materials scientists, and biomedical scientists as well as electrical, chemical, and mechanical engineers, such that great hopes exists of the synergistic effects of interdisciplinarity. Many national initiatives have been launched around the world including Japan, U.S., and European countries, Singapore, Korea, Brazil and South Africa. In the U.S., a report by the National Science and Technology Council claims "The emerging fields of nanoscience and nano-engineering are leading to unprecedented understanding and control over the fundamental building blocks of all physical things. This is likely to change the way that almost everything—from vaccines to computers to automobiles to tires and objects not yet imagined—is designed and made." The National Science Foundation has touted nanoscience as leading to "dramatic changes in the ways materials, devices, and systems are understood and created," and lists among the envisioned breakthroughs "orders-of-magnitude increases in computer efficiency, human organ restoration using engineered tissue, designer materials created from direct assembly of atoms and molecules, and the emergence of entirely new phenomena in chemistry and physics."

How nanotechnology research and development and the marketing of products might reshape and otherwise affect human living conditions as well as beliefs, values, and social relationships is an unanswered question yet. The unknown and potentially substantial harms and benefits, the risks and opportunities it represents to social, cultural, and material life warrants immediate and careful ethical reflection. The effort to engage and develop an ethics for nanotechnology complements other efforts to explore the moral dimensions of the scientific and technological transformations of society, such as environmental ethics (dealing with the impact of human civilization on the natural environment), biomedical ethics (focusing on the growing technological capacities to heal diseases, to enhance human performance, and to control reproduction), and computer ethics (researching the impact on society of the increasing digitalization of information and the computer control of all kinds of processes). And yet there is at least one factor that may at this point distinguish nanotechnology ethics from other areas of engineering ethics. Most nanotechnology pursuits are still in the research, if not visionary, stage and have not emerged as actual development. No one really knows where the initiatives will lead, or what will be the course of nanotechnology research and development. It is touted as having a potentially profound impact on virtually every facet of human life, but that impact is still futuristic, and cannot be predicted with any accuracy. Formulating an ethics of a technology, which has yet to develop, is a daunting quest, to say the least. There are many uncertainties about the moral obligations, which may come as a result of nanotechnology development and which confound the nano-ethics inquiry. To identify moral rules and principles for the future development of nanotechnology, other than traditional rules and principles, requires the impossible task of seeing into the future. However, the importance of doing so is enormous, especially in the engineering classroom where future designers and researchers of nanotechnology are being educated.

Nano-engineering Ethics

At this point, the teaching of nano-engineering ethics is still in a rudimentary and explorative state, since societal and ethical implications of nanotechnology are still a matter of theoretical reflection. Governmental agencies have recently organized several workshops that began stocktaking moral and societal issues and which have encouraged further scholarly reflection. However, it is almost certain that nano-engineering ethics will soon become part of the engineering education, so that it is worth starting to reflect on its implementation now against the background of existing educational approaches in engineering ethics.

Since 2001, all the engineering programs are to be accredited using the Engineering Criteria 2000, which requires, among others, that students should be educated in *professional* and *ethical responsibility* and "the impact of engineering solutions in a global and societal context". Although courses in engineering ethics have been offered long before and in many countries, mostly on a non-mandatory basis, the new requirement has started a debate on educational approaches in engineering ethics. Controversies focus on four different issues:

Curriculum Integration

Besides the question of whether engineering ethics should be mandatory or not, there are different models for integrating it into the general engineering curriculum. The two main issues are whether engineering ethics should be a freestanding course or pervading the engineering curriculum, and whether engineers, philosophers/ethicists, or a team of engineers and philosophers should teach it.

Goals and Topics

Typical goals and topics of engineering ethics include the awareness of moral issues and societal implications of technology, of professional and general moral responsibilities; the learning of moral arguments, judgments, and problem-solving skills; and the understanding of ethical frameworks and the general place of science and technology in society. Different approaches put different emphasis on these topics, however. For instance, philosophers tend to emphasize the learning of ethical theories and skills, whereas engineers tend to focus on particular problem awareness and moral education; US approaches frequently include professional responsibilities to engineering societies, colleagues, and customers, whereas European approaches mostly focus on moral responsibilities to the general society. Another issue, which usually depends on resources, is whether the topics or courses should be specified for different engineering disciplines or be general enough so as to cover all the engineering (and science) disciplines.

Ethics Approach

The old debate in ethical theory (and law), whether moral judgments can and should be based on general principles or particular cases, has, via debates in applied ethics, also pervaded engineering ethics and its education. The *top-down approach* starts teaching ethical theories or professional codes of conduct and deals with particular cases only later as illustrations or derived instances. The *bottom-up approach* first introduces particular cases and problems (taken either from history, forecasting, or fiction) from which other cases are to be developed either by analogy or via general ethical perspectives that first need to be introduced by logical induction. Although, the "bottom-up approach" has become prevailing, because many students understand particular cases better than general principles, it is a matter of debate if students thus achieve general ethical skills to deal also with new problems.

Student Engagement

Depending on the goals of engineering ethics education, there is debate, and room for further innovations, on how students should become engaged, either intellectually or emotionally or both. Besides traditional forms of education, new pedagogic approaches exercise collaborative learning and role-playing or include visits to practicing engineers in companies and institutions.

Much of what is taught in the engineering ethics classroom is a matter of practical, personal decision-making and responsibility—the avoidance of harm and the doing of good in various professional settings. However, in the consideration of newly developing, futuristic technologies such as nanotechnology (and its proposed convergence with biotechnology, and artificial intelligence), there are critical questions to be addressed, but great difficulties in formulating a system of ethical reasoning and guidance from which to pursue those technological developments. How might students consider, for example, the ambition of nanotechnology to "revolutionize" human living through the technological ability to control and manipulate matter with precision? How do we teach the ethics of such a daunting, futuristic claim? How might classroom instruction provide for probing, insightful thinking about this ambition? This paper suggests a "bottom up" approach, which uses science fiction as a tool with which students might navigate the ethical and societal dimensions of such unknown and unfamiliar technological terrain.

Use of Science Fiction

The use of literature for moral education is actually as old as higher education and was implemented in various modern educational models, from the humanistic *Bildung* ideal to the Liberal Arts Education and the Classic Text approach. In fact, the idea that reading their poems, novels, stories, or plays can contribute to the formation of a moral person has been prevailing among writers themselves since early antiquity. Nowadays, literature teachers also use various pieces of classic literature to make students aware of potential moral issues of modern science and engineering, which is one of the existing models of engineering ethics that is frequently ignored in the debates mentioned above. It is questionable, however, whether such 19th-century *mad scientist* stories as Mary Shelley's *Frankenstein*, or the *Modern Prometheus* (1818), or Nathaniel Hawthorne's *Rappaccini's Daughter* (1844) are helpful, since they rather induce anti-scientific cliches based on early modern, if not pre-modern, views of science and thereby reinforce the split of the *two cultures*. Opposed to that, we suggest including selected nanoscience fiction stories in engineering ethics courses. To that end, the stories need to raise moral issues that are considered both important and realistic, in the sense that they are sufficiently complex and that similar scientific and technological capacities are likely to come in the near future.

Within the mentioned debates on educational approaches in engineering ethics, our approach is open to all kinds of *curriculum integration*. The *goals and topics* essentially depend on the selected stories. The approach is, of course, *bottom up*. *Student engagement*, both emotionally and intellectually, is actually one of the strengths as we replace conventional ethics texts with intellectually provocative, entertaining science fiction stories to be discussed in the classroom.

Today, nanotechnology is mostly articulated in visionary terms about industrial revolutions, transformations of society, and radical changes of the human condition. Apart from science fiction authors, many software engineers, scientists, policy makers, business people, and transhumanists are heavily engaged in drafting such grand visions. That is also what the public mainly associates with nanotechnology and what people are mainly fascinated about and afraid of. It has even been argued that nanotechnology is about to blur the boundary between science fiction and non-fiction. Whether all these visions will turn out to be correct or not, scientists and engineers working in nanotechnology research are faced in public discourses with that visionary climate and with great hopes and fears. Whether they like the role or not, nanotechnologists are considered the essential actors of making the greatest dreams and the greatest fears come true. Therefore, more than in any other field, students of nanotechnology must be prepared to respond to such expectations, in public discourse as well as in daily research decisions. They need to understand what these visions are about, what their cultural backgrounds and driving societal forces are.

Since science fiction authors are arguably the most professional and influential vision writers, their texts are an ideal source for making engineering students aware of the public expectations they will increasingly face in their professional life.

Furthermore, engineering students when being confronted with novel phenomena and with unfamiliar, perhaps uncomfortable, curious, and intriguing ideas must have some way of negotiating and responding. Through science fiction, the imaginative process takes the mind beyond recognized material awareness and cognitive boundaries into domains of unexplainable sensory experience and psychological unknowing. It can be used to reach beyond what is known and understood in order to make meaning of those elements of human awareness and perception which otherwise elude one's grasp. It can also help break down conceptual barriers to awareness, and in turn, inspire imaginative consideration of what is right and good in futuristic technological development. In the engineering classroom in particular, science fiction can provide a way for students to cross the barriers of rational limits to knowing, through provocative images that reach past intellectual domains into the collective unconscious of myth, belief, and archetypes towards otherwise inaccessible realities. When notions such as personal transformation emerge in the imagined possibilities and promises of nanoscale manipulations of matter, students need to understand what that may mean in moral terms and how that might change the notion of being persons that underlies all ethical theories.

Science fiction can help students to approach an understanding otherwise inaccessible, except through the realm of intuition, emotion, and imagination. Traditional deontology cannot easily be applied to futuristic technologies if life then may in no way resemble what we now know life to be. To approach ethics of futuristic technologies in the engineering classroom, rationalistic methodologies must be supplemented with awareness of feelings and even fears, excitements and dreams about what students may one day play as either consumers or designers of that future. Only then can they learn to articulate and define moral judgments and standards about the development of technologies that promise such profound changes to the characteristics of human life. Through science fiction, moral imagination can be elicited as a primary analytical tool, so that engineering students can engage and deliberate over moral judgments in light of emotion, attitudes, and preferences, through the creative and illuminative power of the human intuition.

Of course, there are also drawbacks of teaching engineering ethics through science fiction, and we do not recommend to base engineering ethics courses entirely on science fiction. Instead, science fiction provides, for the reasons mentioned above, a useful additional approach in such visionary fields as nanotechnology to be complemented by traditional approaches and careful guidance by teachers. First, students need to be aware of the difference between science fiction and non-fiction, that science fiction is not a prediction of future scenarios but about stimulating the thinking about possible futures, which includes imagining quite different scenarios. Secondly, as with all "bottom-up" approaches, classroom discussion should move from the particularities of the stories to the morally relevant aspects and to the underlying general moral issues, such that students learn general analytical skills of ethics that they can apply to other cases. Supplementary reading of "top-down" approaches might be advisable. Third, while the emotional access to moral issues is an advantage of the science fiction stories, students should also learn to reflect on their own emotions, what part of the stories induced them and on how they influence their moral judgments. Finally, the selection of science fiction stories should be made carefully and balanced to include not only different moral issues but also different perspectives. Students should be provided with critical thinking skills and learn that science fiction authors have their own moral positions and that writing novels can sometimes be a powerful tool to impose moral positions on readers.

Two Examples of Nano-Science Fiction

Some nano-science fiction stories are plainly horrifying, reflecting fears of the unknown, and of the power of misguided science to potentially destroy the foundations of life as we know it. Greg Bear's classic *Blood's Music* and Michael Crichton's recent novel *Prey*, which is currently turned into a movie, are examples. While there is an argument to be made for the importance of the inclusion of such literature in public discourse about developing nanotechnology, its place in the engineering ethics classroom is less clear. We would recommend, instead, writings that more subtly bring to light the ambiguities and complexities of future social and moral life, which are hauntingly plausible under the influence of nanotechnology. Supplementary to the technical education, engineering ethics should make students aware of the complexity of human society, the sensitivity of the environment, and the complex relationship that exists between technology, humans and the environment.

In this section we present two examples of nano-science fiction stories which can be useful in that pedagogical ambition. Both stories are classical texts of nano-science fiction, written between 1989 and 1995. They differ in the kinds of moral and societal questions they pose and in the extent of their visionary horizon. The first story focuses on researchers, their research motives, visions, interactions with other researchers, and their sense of moral responsibility. In the engineering ethics classroom, this text provides a fictional case-based introduction to research ethics by spinning a web of moral conflicts rather by than presenting simple moral answers. The second story develops visions of a technologically radically changed society, most poignantly expressed from a child's perspective. This text is useful to make students reflect on the role of technology in society and its more far-reaching impacts on values and beliefs.

Nanotech Chronicles

In his novel *The Nanotech Chronicles*, Michael Flynn presents his view on the gradual development of future nanotechnology and its social implications in six nano-science fiction stories. The second story, "The Washer at the Ford", is an ideal reading for introducing engineering students to a large variety of moral issues that might arise along with the research and development of new technologies. By masterly arranging six characters with different moral positions, for each of which we find almost convincing arguments, Flynn prompts his readers to reflect on moral issues and to solve moral conflicts.

The main character, Dr. Charles Singer, is described as the inventor of molecular nanotechnology and president of the small American research company Singer Lab. His three collaborators (and associates) are his wife Dr. Jessica Burton-Peeler, the "second best geneticist in North America" and Vice-president of Singer Lab; Dr. Kalpit Patel, a second-generation Indian-American microbiologist and with 52 the oldest of the group; and Eamonn Murchadha, a young Irish microbiologist and "nanomachinist". The group is joined by Dr. Masao Koyanagi, a citizen of the non-allied "codominium of Brazil and Japan", whose "grandparents were killed in Nagasaki from radiation induced cancer", such that the story cannot play in the far future. Based on his former research on radiation resistant bacteria, Koyanagi convinces the group to work on "nanomachines" that should considerably increase the radioactive radiation resistance of humans by genetic modification of corporal cells. Compared to many other nano-science fiction stories, that project does not sound too far-fetched such that there is no barrier for readers to understand the following conflicts. The sixth character in the plot is John Royce, a representative of the American government from the "Office of Technology Assessment", who offers to SingerLab almost unlimited financial support of the project in return for full governmental rights on all results.

As the story develops, all six characters form different positions and interact with each other in a variety of conflicts. Koyanagi has strong idealistic motives to increase the radiation resistance of all humans, across all nationalities and social classes, as this will reduce the human suffering from cancer and other health consequences of radioactivity. Singer is not only a competent scientist, but also an individualistic entrepreneur with strong ambitions to make profits for him and the company, so that he only reluctantly agrees to the governmental contract because his company is in financial problems. Between Koyanagi's responsibility for humanity and Singers focus on his company and himself, Royce engages in moral arguments for the researchers' responsibility to the country and for the national interest. Patel represents the morally disinterested researcher who wants to focus only on technical problems and solutions.

The different moral positions let the first three characters act against each other: Royce wants the foreigner Koyanagi out of the project for which he promises Singer additional money. Singer, in order to secure the governmental funding and because he suspects that Koyanagi's altruism might undermine his profits, is at first willing to fire him. Koyanagi mistrusts both Singer's profit-orientation and the governmental interest and pirates the research results for worldwide dissemination. In the end, Singer and Koyanagi form an alliance against Royce when it turns out that the government is planning to use their research for a secret weapon.

The remaining two characters, Murchadha and Burton-Peeler, raise moral concerns about the potential misuse of the prospective technology and agree that the risks outweigh the benefits. They argue that human radiation resistance would make people careless in dealing with radiation around nuclear power plants, to the detriment of animals and plants as well as of fetus and unprotected humans in case of malfunction. They are particularly concerned that governments might more likely run the risk of a nuclear war because of the illusion of total radiation protection, which in fact would be only limited. While Murchadha is only worried about the dangers, Burton- Peeler is very anxious and sabotages the common project to the dismay of all others, including Murchadha.

Murchadha's and Burton-Peeler' criticism put them in opposition to the other characters. Of course they mistrust Royce and the government. They reproach Singer not only for being blind to the broader societal impacts of the prospective technology, which Burton-Peeler explains to her husband in a longer ethical discourse, but also for being selfish and mercenary. Since they are equally concerned about humanity as Koyanagi, though with contrary positions, the relation between both parties is particularly interesting. Burton-Peeler, the activist who has little trust in the responsibility of humans, sees danger and naivety in Koyanagi and tries to get him out of the project by guiding suspects about her sabotage towards him. Koyanagi, the Buddhist and "bleeding-heart idealist", has "greater faith in human nature" and strongly condemns Burton-Peeler's sabotage.

Without being boring or artificially constructed, Flynn's nano-science fiction story is full of moral issues without undue simplification, each worth discussing at length in an introductory class in engineering ethics. The main issues include:

1. *Dimensions of intended and unintended social consequences of technological innovation*, including attempts to fix unintended consequences by technological implementation, and cultural conservatism: Engineering students see their future activity embedded in a wider social contact, broaden their imagination and learn that good intentions can sometimes result in bad consequences. At the same time, they understand the limits of social foresight and of planning technology-induced social changes.
2. *Intricacy of issues in engineering ethics*, such that even contradictory positions can, each of their own, convincingly appeal to "reason". Students are prompted to solve such contradictions by analyzing for each moral position the different hidden assumptions about the scope of responsibility, fundamental values, and the human nature.
3. *Different kinds of interests and values* that engineers are confronted with in real-case research and development and for which they need to make balanced and responsible decisions: Students learn that, beyond mere technological performance standards, economical, political, and moral values as well as competition, friendship, and social group dynamics actually play important roles in engineering decision-making.
4. *Risk analysis* and the social relativity of risk perception: Students learn risk calculation and its limits in extreme cases (small

probabilities of infinite harm) as well as the difference between objective and subjective risks assessment. They understand that risk perception is not only essentially subjective and dependent on personal hopes and fears, but also embedded in social contexts.

5. *Standard excuses from moral responsibility*, such as technological determinism; higher necessity; technological naivety; the appeal to good intentions. By reading such excuses and their responses in moral debates throughout the story, students learn the weakness and narrow-mindedness of such phrases and how to replace them with better arguments.

"The Diamond Age"

While telling the story of a little girl named Nell, Stephenson's novel *The Diamond Age* weaves the reader through a world that is at once familiar and strange. Much as with today, in that world, human beings have divided themselves into rather isolated societies, living in self-ruling territories and being determined primarily by racial identity and common ideas and rituals. Wars are fought over material resources, land, and power. Criminals are judged according to the severity of their deeds, and entertainment is accorded significant value. However, unlike today's world, in Stephenson's fictional world, nanotechnology is able to provide for all of the basic human needs, and many other material desires. "Matter Compilers" use "feed" to build from molecules any object that is programmed into the system, from food to clothing to entire buildings. Scarcity of resources is a political maneuver by powerful territories to both limit and control the use and possible abuse of technological power. This is one reason war persists, such as in the "Chinese Civil War" where people in the North have no access to nanotechnology.

Stephenson's world combines highly developed technology with traditional, sometimes archaic, social conditions. In some territories, social justice is suppressed under the pursuit of information acquisition. There are world governances, but the independence of each territory leaves social conditions up to the individual governments, so that in the "Chinese Coastal Republic" the old ("Confucian") laws prevail and criminals are sometimes punished with physical torture or death. In one such occurrence, a judge from "New China" orders a magistrate to "revert to the time-honored methods of his venerable predecessors." The prisoner is strapped to a heavy metal rack that was normally used for canings, stripped from waist down and situated over a bucket for elimination. Earlier, the court physician "thrusts in a spinal tap," introducing a set of "nanotechnological parasites". These parasites had "migrated up and down the prisoner's spinal column" through the cerebrospinal fluid, and "situated themselves on whatever afferent nerves they happened to bump up against. These nerves, used by the body to transmit information such as (to name only one example), excruciating pain to the brain, had a distinctive texture and appearance that the 'sites were clever enough to recognize..." and the ability to "transmit bogus information alone those nerves". Despite their extraordinary technological abilities, the inhabitants of this future world are troubled, competing, and perpetually in struggle.

The story's main character, a young girl named Nell, is abused, neglected, and impoverished. As Stephenson's narration gives the reader glimpses of many different characters, their world and their struggles, it is Nell's personal journey out of the hell of her own life and into a state of enlightened freedom that consumes the reader's moral hopefulness in an otherwise dark novel. And that hope is embedded behind a "book" called the Primer. Books are not only obsolete, they are forbidden. Information is transmitted only through "nano-electronics". An illegally developed device, because it would threat the established social structure, the Primer is read, like a book, but responds in real time to the fantasies and desires, fears, and imaginings of its reader. The Primer generates the story as the reader proceeds to read it by responding to the queries the reader poses to it. Nell is particularly vulnerable to the book's responsive writing, as it senses her physical poverty and her emotional needs. In due time, the book begins to "know" her better than she knows herself and becomes her companion, confidently and ultimately her savior in the face of real threats to her life.

More than just an inexhaustible source of information, the Primer is a superintelligent computing device that interacts with its user in a quasi-personal manner. It is omniscient and perfectly empathizing, prudent, foresighted, and benevolent—a device that meets various human desires through advanced information technology. Retrospectively, the psychological and social conditions have drastically changed with each innovation in information technology, from the inventions of speaking, writing, and printing to digital computers. In *The Diamond Age*, Stephenson offers a fantastical but early plausible glimpse at what it may mean to be alive in human society human when nanotechnology reshapes information technology once more. There are many significant strands of social-ethical inquiry in *The Diamond Age* that might be considered in an engineering ethics classroom, including:

Social reality of technological visions

Various proponents of the nanotechnology "revolution" have suggested that by manipulating atoms with precision, humans will have more control, and through that, fewer material needs and eventually less struggle. Stephenson's story reconsiders the premise that material wants and needs are the basis for human struggle. In *The Diamond Age* social and personal struggle persist, as does material need, despite the highly developed capacities of nanotechnology. Human struggle is depicted in this story not simply as an outgrowth of material want, but from more complex elements of the human psyche and social structure.

Concept of privacy

In Stephenson's story the concept has lost any meaning because new technologies radically have changed expectations and behaviours regarding the information humans extrude from one another. In one scene, the narrator explains that the foyer of a home has come to be used in new ways, because all visitors were carefully examined before entering into a home's "inner sanctum" ... ("So elaborate waiting room etiquette had flourished, and sophisticated people all over the world understood that when they called upon someone, even a close friend, they could expect to spend some time sipping tea and pursuing magazines in front of a room infested with unobtrusive surveillance equipment"). One of the stated interests of today's governments in nanotechnology is the development of undetectable surveillance systems to be placed in a variety of locations and materials. In *The Diamond Age*, that technological ability is moved beyond military or national security use into daily use by individuals, organizations, and businesses. Engineering students might consider the values embedded in the information gathering devices being designed today, considering that in the future, "...when everything can be surveyed, all we have left is politeness".

Nature and technology

In Stephenson's futuristic world, nature, in the sense of both the natural environment and the human body, is reduced to mere instrumental objects. Birth control, for example, is pursued through a technology called "the Freedom Machine" where "mites" are placed inside the body of the female to eat fertile eggs as they are formed. Excessive sexual promiscuity is the norm, due to the "freedom" from conceptualization and the ability to intercept sexually transmitted diseases. Trees have no other use than for the production of oxygen, which could also be done by nanotechnology, so that they are removed for unlimited land development. An interesting discussion subject for the engineering ethics classroom is the relationship between nature and technology, which in Stephenson's science fiction is used to master and control otherwise natural processes. What values are driving the instrumentalization and "mastery" of nature and what social and moral consequences might this process have?

Conclusion

Through science fiction, engineering students are given opportunities to move beyond ideas of present material reality into the domains of the imagined future, where they can work with moral questions of our future with nanotechnology in creative and active ways. From their engagement with science fiction, they can move back to real time and the actual state of nanotechnology research and development, to ask where we may possibly move with these technologies, and how they may affect social, moral, and environmental conditions of human life. Most importantly, the imaginative process can better equip them to engage reflection over such issues as what we take to be our most cherished values and beliefs, and how those values and beliefs might drive the technology development and in turn be impacted by the technologies we are creating. By bringing this imaginative process of reflection into the engineering classroom, engineering students are freed from the constraints of technical problem solving efforts to reflect more deeply upon the ethical dimensions of the emerging nanotechnology age.

If accompanied by guided classroom discussion, selected science fiction stories can be used to introduce students to pertinent ethical concepts, to train awareness of moral issues, to educate skills for solving moral conflicts, and to prepare them for their future role as responsible engineers. In addition, students of nano-engineering can learn from classical nano-science fiction novels something about the visionary roots of their field, the public image of nanotechnology, and the great hopes and fears that they as professionals will be faced with.

INDEX